THE PATTERN ON THE STONE

THE PATTERN ON
THE STONE

....

The Simple Ideas That
Make Computers Work

W. DANIEL HILLIS

BOOKS

A Member of the
Perseus Books Group

Grateful acknowledgment is made for permission from the following:

Illustration of the macaque visual cortex from "Distributed Hierarchical Processing in Primate Cerebral Cortex," *Cerebral Cortex* 1: 1–47, courtesy of Oxford University Press.

Pablo Picasso's *Bernard Picasso with His Mother* (1959); © 1998 Estate of Pablo Picasso/Artist Rights Society (ARS), New York.

Photo of the Tinker Toy computer courtesy of The Computer Museum.

The Science Masters Series is a global publishing venture consisting of original science books written by leading scientists and published by a worldwide team of twenty-six publishers assembled by John Brockman. The series was conceived by Anthony Cheetham of Orion Publishers and John Brockman of Brockman, Inc., a New York literary agency, and developed in coordination with Basic Books.

• • • • • • • • • • • • • •

The Science Masters name and marks are owned by and licensed to the publisher by Brockman Inc.

• • • • • • • • • • • • • •

• • • • • • • • • • • • • •

Published by Basic Books,
A Member of the Perseus Books Group

• • • • • • • • • • • • • •

• • • • • • • • • • • • • •

FIRST EDITION

• • • • • • • • • • • • • •

Illustrations by Patricia Isaacs, Parrot Graphics

• • • • • • • • • • • • • •

Library of Congress Cataloging-in-Publication Data
Hillis, W. Daniel.
   The pattern on the stone : the simple ideas that make computers
work / W. Daniel Hillis.—1st ed.
      p.   cm.—(Science masters)
   Includes index.
   ISBN 0-465-02595-1
   1. Computers.   I. Title.   II. Series: Science masters series.
QA76.5.H4918   1998
004—dc21   98-38888
                                                                      CIP

• • • • • • • • • • • • • •

   99 00 01 02 ❖/RRD 10 9 8 7 6

CONTENTS

To Marvin Minsky,
teacher and friend

PREFACE: MAGIC IN THE STONE

I etch a pattern of geometric shapes onto a stone. To the uninitiated, the shapes look mysterious and complex, but I know that when arranged correctly they will give the stone a special power, enabling it to respond to incantations in a language no human being has ever spoken. I will ask the stone questions in this language, and it will answer by showing me a vision: a world created by my spell, a world imagined within the pattern on the stone.

A few hundred years ago in my native New England, an accurate description of my occupation would have gotten me burned at the stake. Yet my work involves no witchcraft; I design and program computers. The stone is a wafer of silicon, and the incantations are software. The patterns etched on the chip and the programs that instruct the computer may look complicated and mysterious, but they are generated according to a few basic principles that are easily explained.

Computers are the most complex objects we human beings have ever created, but in a fundamental sense they are remarkably simple. Working with teams of only a few dozen people, I have designed and built computers containing billions of active parts. The wiring diagram of one of these machines, if it were ever to be drawn, would fill all the books in a good-sized public library, and nobody would have the patience to scan the whole of it. Fortunately, such a diagram is unnecessary, because of the regularity of a computer's design. Computers are built up in a hierarchy of parts, with each part repeated many times over. All you need to understand a computer is an understanding of this hierarchy.

Another principle that makes computers easy to understand is the nature of the interactions among the parts. These interactions are simple and well-defined. They are also usually one-directional, so that the actions of the computer can be sorted neatly into causes and effects, making the inner workings of a computer more comprehensible than, say, the inner workings of an automobile engine or a radio. A computer has a lot more parts than a car or a radio does, but it's much simpler in the way the parts work together. A computer is not dependent so much on technology as on ideas.

Moreover, the ideas have almost nothing to do with the electronics out of which computers are built. Present-day computers are built of transistors and wires, but they could just as well be built, according to the same principles, from valves and water pipes, or from sticks and strings. The principles are the essence of what makes a computer compute. One of the most remarkable things about computers is that their essential nature transcends technology. That nature is what this book is about.

This is the book I wish I had read when I first started learning about the field of computing. Unlike most books on computers—which are either about how to use them or about the technology out of which they're built (ROM, RAM, disk drives, and so on)—this is a book about *ideas*. It explains, or at least introduces, most of the important ideas in the field of computer science, including Boolean logic, finite-state machines, programming languages, compilers and interpreters, Turing universality, information theory, algorithms and algorithmic complexity, heuristics, uncomutable functions, parallel computing, quantum computing, neural networks, machine learning, and self-organizing systems. Anyone interested enough in computers to be reading this book will probably have encountered many of these ideas before, but outside of a formal education in computer science there are few opportunities to see how they all fit together. This book makes the connections—all the way from simple physical processes like the closing of a switch to the learning and adaptation exhibited by self-organizing parallel computers.

A few general themes underlie an exposition of the nature of computers: the first is the principle of *functional abstraction,* which leads to the aforementioned hierarchy of causes and effects. The structure of the computer is an example of the application of this principle—over and over again, at many levels. Computers are understandable because you can focus on what is happening at one level of the hierarchy without worrying about the details of what goes on at the lower levels. Functional abstraction is what decouples the ideas from the technology.

The second unifying theme is the principle of the *universal computer*—the idea that there is really only one kind of computer, or, more precisely, that all kinds of computers are alike in what they can and cannot do. As near as we can tell, any computing device, whether it's built of transistors, sticks and strings, or neurons, can be simulated by a universal computer. This is a remarkable hypothesis: as I will explain, it suggests that making a computer think like a brain is just a matter of programming it correctly.

The third theme in this book, which won't be fully addressed until the last chapter, is in some sense the antithesis of the first. There may be an entirely new way of designing and programming computers—a way not based on the standard methods of engineering. This would be exciting, because the way we normally design systems begins to break down when the systems become too complicated. The very principles that enable us to design computers lead ultimately to a certain fragility and inefficiency. This weakness has nothing to do with any fundamental limitations of information-processing machines—it's a limitation of the hierarchical method of design. But what if instead we were to use a design process analogous to biological evolution—that is, a process in which the behaviors of the system *emerge* from the accumulation of many simple interactions, without any "top-down" control? A computing device designed by such an evolutionary process might exhibit some of the robustness and flexibility of a biological organism—at least, that's the hope. This approach is not yet well understood, and it may turn out to be impractical. It is the topic of my current research.

In an explanation of the nature of computers, there are some fundamentals that have to be dealt with before we can move on to the good stuff. The first two chapters introduce the fundamentals: Boolean logic, bits, and finite-state machines. The payoff is that by the end of chapter 3 you'll understand how computers work, top to bottom. This sets the stage for the exciting ideas about universal computing machines, which begin in chapter 4.

The philosopher Gregory Bateson once defined information as "the difference that makes a difference." Another way of saying this is that information is in the distinctions we choose to make significant. In a primitive electrical calculator, say, information is indicated by light bulbs that go on or off depending on whether a current is flowing or not. The voltage of the signal doesn't matter, nor does the direction of current flow. All that matters is that a wire carries one of two possible signals, one of which causes a bulb to light. The distinction that we choose to make significant—the difference that makes a difference, in Bateson's phrase—is between current flowing and not flowing. Bateson's definition is a good one, but the phrase has always meant something more to me. In my lifetime of four decades, the world has been transformed. Most of the changes we've seen in business, politics, science, and philosophy in that time have been caused by, or enabled by, developments in information technology. A lot of things are different in the world today, but the difference that has made the difference has been computers.

These days, computers are popularly thought of as multi-media devices, capable of incorporating and combining all previous forms of media—text, graphics, moving pictures, sound. I think this point of view leads to an underestimation of the computer's potential. It is certainly true that a computer can incorporate and manipulate all other media, but the true power of the computer is that it is capable of manipulating not just the *expression* of ideas but also the ideas themselves. The amazing thing to me is not that a computer can hold the contents of all the books in a library but that it can notice relationships between the concepts described in the books—not that it can display a picture of a bird in flight or a galaxy spinning but that it can imagine and predict the consequences of the physical laws that create these wonders. The computer is not just an advanced calculator or camera or paintbrush; rather, it is a device that accelerates and extends our processes of thought. It is an imagination machine, which starts with the ideas we put into it and takes them farther than we ever could have taken them on our own.

....

CHAPTER I

NUTS AND BOLTS

When I was a child, I read a story about a boy who built a robot out of parts he found lying around a junkyard. The boy's robot could move, talk, and think, just like a person, and it became his friend. For some reason, I found the idea of building a robot very appealing, so I decided to build one myself. I remember collecting body parts—tubes for the arms and legs, motors for the muscles, lightbulbs for the eyes, and a big paint can for the head—in the full and optimistic expectation that after they were assembled and the contraption was plugged in, I would end up with a working mechanical man.

After nearly electrocuting myself a few times, I began to get my parts to move, light up, and make noises. I felt I was making progress. I began to understand how to construct movable joints for the arms and legs. But something even more important was beginning to dawn on me: I didn't have the slightest idea how to control the motors and the lights, and I realized that something was missing in my knowledge of how robots worked. I now have a name for what was missing: it's called *computation*. Back then, I called it "thinking," and I saw that I didn't have a clue about how to get something to think. It seems obvious to me now that computation is the hardest part of building a mechanical man, but as a child this came as a surprise.

BOOLEAN LOGIC
·····················

Fortunately, the first book I ever read on the subject of computation was a classic. My father was an epidemiologist, and we were living in Calcutta at the time. Books in English were hard to come by, but in the library of the British consulate I found a dusty copy of a book written by the nineteenth-century logician George Boole. The title of the book was what attracted me: *An Investigation of the Laws of Thought.* This grabbed my imagination. Could there really be laws that governed thought? In the book, Boole tried to reduce the logic of human thought to mathematical operations. Although he did not really explain human thinking, Boole demonstrated the surprising power and generality of a few simple types of logical operations. He invented a language for describing and manipulating logical statements and determining whether or not they are true. The language is now called *Boolean algebra.*

Boolean algebra is similar to the algebra you learned in high school, except that the variables in the equations represent logic statements instead of numbers. Boole's variables stand for propositions that are either true or false, and the symbols $\wedge$, $\vee$, and $\neg$ represent the logical operations **And, Or,** and **Not**. For example, the following is a Boolean algebraic equation

$$\neg(A \vee B) = (\neg A) \wedge (\neg B)$$

This particular equation, called De Morgan's theorem (after Boole's colleague Augustus De Morgan), says that if neither A nor B is true, then both A and B must be false. The variables A and B can represent any logical (that is, true or false) statement. This particular equation is obviously correct, but Boolean algebra also allows much more complex logical statements to be written down and proved or disproved.

Boole's work found its way into computer science through the master's thesis of a young engineering student at the Massachusetts Institute of Technology named Claude Shannon. Shannon is best known for having invented a branch of mathematics called *information theory*, which defines the measure of information we call a *bit*. Inventing the bit was an impressive accomplishment, but what Shannon did with Boolean logic was at least as important to the science of computation. With these two pieces of work, Shannon laid the foundation for the developments that were to occur in the field of computing for the next fifty years.

Shannon was interested in building a machine that could play chess—and more generally in building mechanisms that imitated thought. In 1940, he published his master's thesis, which was titled "A Symbolic Analysis of Relay Switching Circuits." In it, he showed that it was possible to build electrical circuits equivalent to expressions in Boolean algebra. In Shannon's circuits, switches that were open or closed corresponded to logical variables of Boolean algebra that were true or false. Shannon demonstrated a way of converting any expression in Boolean algebra into an arrangement of switches. The circuit would establish a connection if the statement was true and break the connection if it was false. The implication of this construction is that any function capable of being described as a precise logical statement can be implemented by an analogous system of switches.

Rather than presenting the detailed formalisms developed by Boole and Shannon, I will give an example of their application in the design of a very simple kind of computing device, a machine that plays the game of tic-tac-toe. This machine is much simpler than a general-purpose computer, but it demonstrates two principles that are important in any type of computer. It shows how a task can be reduced to *logical functions* and how such functions can be implemented as a circuit of connected switches. I actually built a tic-tac-toe machine out of lights and switches shortly after I read Boole's book in Calcutta, and this was my introduction to computer logic. Later, when I was an undergraduate at MIT, Claude Shannon became a friend and teacher, and I discovered that he, too, had used lights and switches to build a machine that could play tic-tac-toe.

As most readers know, the game is played on a 3 x 3 square grid. Players take turns marking the squares, one player using an *X*, the other an *O*. The first player to place three symbols in a row (horizontally, vertically, or diagonally) wins the game. Young children enjoy tic-tac-toe because it seems to offer limitless possible strategies for winning. Eventually they realize that only a small number of patterns can occur, and the game consequently loses its charm: once both players learn the patterns, each game invariably ends in a tie. Tic-tac-toe is a good example of a computation precisely because it wavers on this line between the complex and the simple. Crossing that line is what computation is all about. Computation is about performing tasks that seem to be complex (like winning a game of tic-tac-toe) by breaking them down into simple operations (like closing a switch).

In tic-tac-toe, the situations that occur are few enough so that it's practical to write them all down, and therefore to build the correct response in every case into the machine. We can use a simple two-step process for designing the machine: *first,* reduce the play to a series of cases defining the correct response to each pattern of moves; *second,* convert those cases into electrical circuits by wiring the switches to recognize the pattern and indicate the appropriate response.

One way to proceed would be to write down every conceivable arrangment of *X's* and *O's* which could be placed on the grid and then decide how the computer would play in each instance. Since each of the nine squares has three possible states (*X*, *O*, and blank), there are 3^9 (or 19,683) ways to

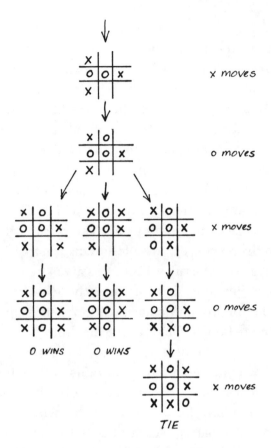

FIGURE I

Part of a game tree for tic-tac-toe

fill the grid. But most of these patterns would never occur in the course of a game. A better method of listing the possibilities is to draw up a *game tree*—a configuration that traces every possible line of play. The game tree starts with a blank grid at the root and has a branch for every possible alternative line of play, determined by the move of the human player. (The tree does not need to branch when the machine plays, because the response of the machine to any given move is always predetermined.) Figure 1 shows a small part of such a tree. For every possible move made by *X*, the human player, there is a predetermined *O* response to be made by the machine. (For some strange reason, computer scientists always draw trees upside-down, with the "root" at the top.)

The tree in Figure 1 illustrates the strategy that I always use in tic-tac-toe: I play in the center whenever I can. The machine's moves are determined by the human player's moves, which vastly reduces the number of possibilities to be considered. A full game tree, showing what the machine should do in every situation, has about five hundred or six hundred branches, the exact number depending on the details of strategy. Following the tree will cause the machine to win, or at least tie, every game. The rules of the game are built into the responses, so by following the tree the machine will always obey the rules. From this game tree, we can write down specifications that say exactly when the machine should play in any particular position. These specifications constitute the Boolean logic of the machine.

Once we have defined the desired behavior, we can translate that behavior into electrical circuits built out of batteries, wires, switches, and lights. The basic circuit in the machine is the same circuit used in a flashlight: when the switch is pressed down—that is, closed—the light goes on, because a complete path has been formed between the bulb and the battery. (The connections to the battery are indicated by the + and − signs.) Most important, these switches can be wired either *in series* or *in parallel*. For instance, we can put two switches together in series to make a light that works only when both switches are closed. This circuit implements one of the basic switching functions of the computer—the "logic block" known as the **And** function, so called because the bulb lights only when the first **and** the second switches are closed. Switches connected in parallel form the **Or** function, which connects the circuit (and thus lights the bulb) whenever either **or** both of the switches are closed (see Figure 2).

These simple patterns of serial and parallel wiring can be used in combinations to form connections that follow various logical rules. In the tic-tac-toe machine, chains of

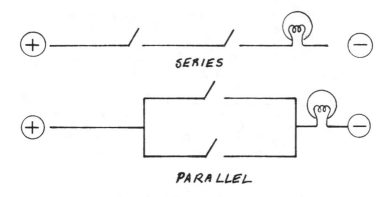

FIGURE 2

Switches in series and parallel

switches connected in *series* are used to detect patterns, and
these chains are connected in *parallel* to lights, so that several patterns can light the same bulb—that is, produce the
same response from the machine.

The tic-tac-toe machine I built has four banks of nine
switches each, and each switch corresponds to one of the nine
squares on the tic-tac-toe grid. It also has nine lightbulbs,
arranged in the pattern of a tic-tac-toe board. The machine,
which always plays first, makes its moves by lighting a bulb.
The human player moves by closing a switch—using the first
bank of switches to make his first move, the second bank for
his second move, and so on. In my version, the machine
always begins by playing in the upper left corner of the board,
a scheme that reduces the number of cases considerably. The
human player responds by closing one of the switches in the
first bank (say, the one corresponding to the center square in
the grid), and the game proceeds. The machine's strategy is
embodied in the wiring between the switches and the lights.

The wiring that produces the machine's first response is
easy (see Figure 3). Each switch in the first bank is connected to a light that corresponds to the machine's reply. For
instance, a play in the center causes a response in the lower

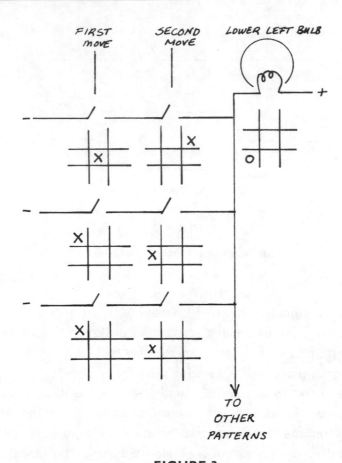

FIGURE 3

Several different patterns that produce the same response

right, so the center switch is wired to the lower-right light. Since my machine always responds in the center square if it can, most of the first bank of switches is wired in parallel to the middle light.

Each pattern for the second round of play depends on the human player's first and second moves. To recognize this combination of human moves, the corresponding switches are wired in series. For example, if the player's first move is in the center and second move in the upper right, the machine is then supposed to respond by playing in the lower left. This pattern is accomplished by wiring the center switch in the first bank in series with the upper-right switch in the second bank ("if center **and** upper-right squares are filled, then . . . "), with the chain of two switches being connected to the lightbulb in the lower left. Each parallel connection to a bulb specifies a different combination that will cause the bulb to light ("this move or that move will provoke this response"). Whenever it was necessary to use the same switch in two different circuits, I used a "double throw" switch—two switches mechanically linked to the same button, so that they switch together—which allows the same move to be part of two different patterns. The wiring of the third and fourth banks of switches follows the same principle, but there are even more combinations. As you can imagine, the wiring gets complicated, even though the principles are simple. There are fewer choices open on the grid, but the chains of switches are longer.

The tic-tac-toe machine I built has about a hundred and fifty switches. This seemed like a lot to me at the time (I made the switches out of wood and nails), but the computer chips I design today have millions of switches, most of them connected in patterns very similar to those used in the tic-tac-toe machine. Most modern computers use a different kind of electrical switch—a transistor, which I will describe later—but the basic notion of connecting switches in series to produce the **And** function and connecting switches in parallel to produce the **Or** function is exactly the same.

While the logic of the tic-tac-toe machine is similar to the logic of a computer, there are several important differences. One is that the tic-tac-toe machine has no notion of events happening sequentially in time; therefore, the entire sequence of the game—that is, the entire game tree—must be determined in advance. This is cumbersome enough where tic-tac-toe is concerned and practically impossible for a more complicated game, like chess, or even checkers. Modern computers are very good at playing checkers and pretty good at playing chess (see chapter 5), because in place of the pre-determined game tree they use a different method—one that involves examining patterns sequentially in time.

Another difference between the tic-tac-toe machine and a general-purpose computer is that the tic-tac-toe machine can perform only one function. The "program" of the machine is built into its wiring. The tic-tac-toe machine has no software.

BITS AND LOGIC BLOCKS

As I noted in the Introduction, there is no reason the tic-tac-toe machine (or any other computer) has to be built out of electrical switches. A computer can represent information using electrical currents, fluid pressures, or even chemical reactions. Whether you build a computer out of transistors, hydraulic valves, or a chemistry set, the principles on which it operates are much the same. The key idea of the tic-tac-toe machine is that the **And** function is implemented by connecting two switches in series and the **Or** function is implemented by connecting two switches in parallel, but there are many other ways to implement **And** and **Or**.

Here I must pause to mention the *bit.* The smallest "difference that makes a difference" (to use Bateson's phrase again) is a difference that splits all signals into two distinct classes. In the tic-tac-toe machine, the two classes are "current flowing" and "no current flowing." By convention, we call the two possible classes 1 and 0. These are just names; we could as easily call them **True** and **False**, or **Alice** and **Bob**. Even the choice of which class is called 0 and which is called 1 is arbitrary. A signal that can carry one of two different messages (like 1 or 0) is called a *binary* signal, or a *bit.* A computer uses combinations of bits to represent all kinds of sets of alternatives—different moves in tic-tac-toe, say, or different colors to be displayed on a screen. Since the conven-

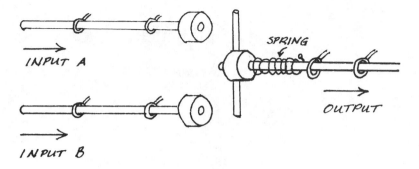

FIGURE 4

Mechanical implementation of the OR function

tion is to designate the bits by 1's and 0's, people often think of these bit patterns as numbers, hence the old chestnut "The computer does everything with numbers." But this convention is simply a way of thinking about what's going on. If we had named the two possible messages conveyed by the bit the letters X and Y, people would be saying, "The computer does everything with letters." The more accurate statement is "The computer represents numbers, letters, and everything else with patterns of bits."

Instead of using the flow of electricity to represent a bit, we could have used mechanical motion. Figure 4 shows how the **Or** function is implemented using a technology that represents 1 by sliding a stick to the right. As long as both the *A* and the *B* input sticks stay to the left, representing 0, then the spring will keep the output stick pushed to the left, but if either input stick slides to the right, then the output stick will slide to the right also. The object in Figure 5 computes another useful function, that of inversion: The inverter turns every signal into its opposite: for example, it turns a push to the right into a pull to the left, and vice versa.

These **And**, **Or**, and **Invert** functions are *logic blocks*, and they can be connected in order to create other functions. For instance, the output of an **Or** block can be connected to an **Invert** block to create a **Nor** function: the **Nor**

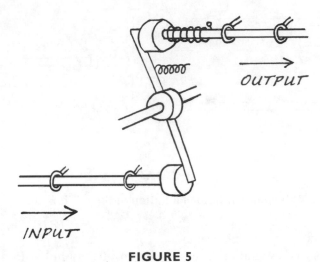

FIGURE 5

Mechanical inverter

output will be a 1 when neither of its inputs is 1. In another example (using De Morgan's theorem), we can make an **And** block by connecting two **Invert** blocks to the inputs of an **Or** block and connecting a third **Invert** block to the output (see Figure 6). These four work together to implement the **And** function, so the final output is 1 only when both the inputs are 1.

Early computing devices were made with mechanical components. In the seventeenth century, Blaise Pascal built a mechanical adding machine, which inspired both Gottfried Wilhelm Leibniz and the English polymath Robert Hooke to build improved machines that could multiply, divide, and even take square roots. These machines were not programmable, but in 1833 another Englishman, the mathematician and inventor Charles Babbage, designed and partially constructed a programmable mechanical computer. Even as late as my own childhood in the sixties, most arithmetic calculators were mechanical. I've always liked these mechanical machines, because you can see what's happening, which is not the case with electronic

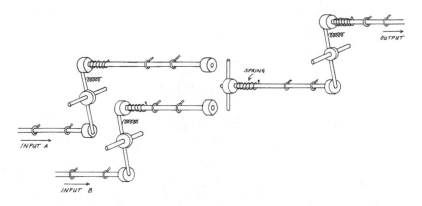

FIGURE 6

An And **block constructed by connecting an** Or **block to inverters**

computers. When I'm designing an electronic computer chip, I imagine the operation of the circuits as moving mechanical parts.

THE FLUID COMPUTER

The picture I have in my mind when I design a logic circuit is of hydraulic valves. A hydraulic valve is like a switch that controls and is controlled by the flow of water. Each valve has three connections: the input, the output, and the control. Pressure on the control connection pushes on a piston that turns off the water flow from input to output. Figure 7 shows a circuit for the **Or** function, built out of hydraulic valves.

In this circuit, water pressure is used to distinguish between the two possible signals. Notice that in a hydraulic valve the control pipe can affect the output pipe but the output pipe cannot affect the control pipe. This restriction establishes a forward flow of information through the switch; in a sense, it establishes a direction in time. Also, since the valve is

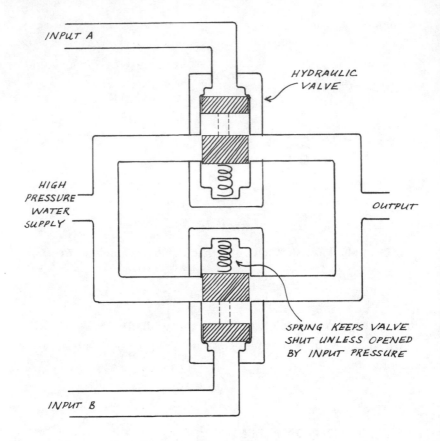

FIGURE 7

An Or block built with hydraulic valves

either open or closed, it serves an additional function of *amplification*, which allows the strength of the signal to be restored to its maximum value at every stage. Even if the input is a little low on pressure—because it goes through a long, thin pipe, say, or because of a leak—the output will always be at full pressure thanks to the on/off operation of the valve. This is the fundamental difference between *digital* and *analog*: A digital valve is either on or off; an analog valve, like your kitchen faucet, can be anything in between. In the hydraulic computer, all that is required of the input signal is that it be strong enough to move the valve. In this case, the difference that makes a difference is the difference in water pressure sufficient to switch the valve on. And since a weakened signal entering an input will still produce a full-strength output, we can connect thousands of layers of logic, the output of one layer controlling the next, without worrying about a gradual decrease in pressure. The output of each gate will always be at full pressure.

This type of design is called *restoring logic*, and the example in hydraulic technology is particularly interesting, because it corresponds almost exactly to the logic used in modern electronic computers. The water pressure in the pipes is analogous to the voltage on the wires, and the hydraulic valve is analogous to the metal-oxide transistor. The control, input, and output connections on the valve correspond closely to the three connections (called *gate, source,* and *drain*) on a transistor. The analogy between water valves and transistors is so exact that you could translate the design for a modern microprocessor directly into a design for a hydraulic computer. To do so, you would need to look at the pattern of wires on the silicon chip under a microscope and then bend a set of pipes into the same shapes as the wires on the chip and connect them in exactly the same pattern. In place of each transistor, you would use a hydraulic valve. The pipe that corresponds to the power-supply voltage on your chip would be connected to a pressurized water supply, and the pipe that corresponds to the ground connection could empty down a drain.

To use the hydraulic computer, you would have to connect hydraulic equivalents of its inputs and outputs—you would need to build a hydraulic keyboard, a hydraulic display, hydraulic memory chips, and so on—but if you did all this, it would go through exactly the same switching events as the electronic chip. Of course, the hydraulic computer would be much slower than your latest microprocessor (to say nothing of larger), because water pressure travels down pipes much more slowly than electricity travels down wires. As to the size: Since the modern microchip has several million transistors, its hydraulic equivalent would require several million valves. A transistor in a chip is about a millionth of a meter across; a hydraulic valve is about 10 centimeters on a side. If the pipes scale proportionally, then the hydraulic computer would cover about a square kilometer with pipes and valves. From an airplane, it would look roughly the same as the electronic chip does under a microscope.

When I design a computer chip, I draw lines on a computer screen, and the pattern is reduced (in a process analogous to photographic reduction) and etched onto a chip of silicon. The lines on the screen are my pipes and valves. Actually, most computer designers don't even bother drawing lines; instead, they specify the connections between **And**s and **Or**s and let a computer work out the details of placement and geometry of the switches. Most of time, they forget about the technology and concentrate on the function. I do this, too, sometimes, but I still prefer to draw my own shapes. Whenever I design a chip, the first thing I want to do is look at it under a microscope—not because I think I can learn something new by looking at it but because I am always fascinated by how a pattern can create reality.

TINKER TOYS
··················

Except for the miracle of reduction, there is no special reason to build computers with silicon technology. Building a computer out of any technology requires a large supply of only two kinds of elements: *switches* and *connectors*. The switch is a steering element (the hydraulic valve, or the transistor), which can combine multiple signals into a single signal. Ideally, the switch should be asymmetrical, so that the input signal affects the output signal but not vice versa, and it should have a restoring quality, so that a weak or degraded

FIGURE 8

Tinker Toy computer

input signal will not result in a degraded output. The second element, the connector, is the wire or pipe that carries a signal between switches. This connecting element must have the ability to branch, so that a single output can feed many inputs. These are the only two elements necessary to build a computer. Later we will introduce one more element—a register, for storing information—but this can be constructed of the same steering and connecting components.

I have never built a hydraulic computer, but once, with some friends, I did construct a computer out of sticks and strings. The pieces came from a children's construction set called Tinker Toys. Readers may remember this as a set of cylindrical wooden sticks that fit into fat little wooden hubs with holes in them. The logic of my Tinker Toy computer worked much like that shown in Figure 8. Like the switches-and-lights computer, the Tinker Toy computer played tic-tac-toe. It never lost. The computer was a lot of trouble to make, requiring tens of thousands of pieces from more than a hundred Tinker Toy "Giant Engineer" construction sets, and the finished product (now sitting in the Computer Museum in Boston, Massachusetts) looks incomprehensibly complex. Yet the principles on which it operates are just the simple combination of **And** and **Or** functions described above.

The big mistake I made in designing the Tinker Toy computer is that I did not use *restoring logic*—that is, there was no amplification from one stage of logic to the next. The implementation of the logic was based on sticks pressing against sticks, in a design similar to the one illustrated in figure 4. Because of this design choice, all the force required to move the hundreds of elements in the machine had to be supplied by the press of the input switch. The accumulated force tended to stretch the strings that transmitted the motion, and because there was no restoration at each stage, the errors caused by the stretching accumulated from one logic element to the next. Unless the strings were constantly tuned, the machine would make mistakes.

I constructed a later version of the Tinker Toy computer which fixed the problem, but I never forgot the lesson of that first machine: the implementation technology must produce perfect outputs from imperfect inputs, nipping small errors in the bud. This is the essence of digital technology, which restores signals to near perfection at every stage. It is the only way we know—at least, so far—for keeping a complicated system under control.

FREE TO WORRY ABOUT THE DIFFERENCE THAT MAKES A DIFFERENCE

Naming the two signals in computer logic 0 and 1 is an example of functional abstraction. It lets us manipulate information without worrying about the details of its underlying representation. Once we figure out how to accomplish a given function, we can put the mechanism inside a "black box," or a "building block" and stop thinking about it. The function embodied by the building block can be used over and over, without reference to the details of what's inside. This process of functional abstraction is a fundamental in computer design—not the only way to design complicated systems but the most common way (later, I'll describe an alternate method). Computers are built up of a hierarchy of such functional abstractions, each one embodied in a building block. The blocks that perform functions are hooked together to implement more complex functions, and these collections of blocks in turn become the new building blocks for the next level.

This hierarchical structure of abstraction is our most powerful tool in understanding complex systems, because it lets us focus on a single aspect of a problem at a time. For instance, we can talk about Boolean functions like **And** and **Or** in the abstract, without worrying about whether they are built out of electrical switches or sticks and strings or water-operated valves. For most purposes, we can forget about technology. This is wonderful, because it means that almost everything we say about computers will be true even when transistors and silicon chips become obsolete.

CHAPTER 2

··

UNIVERSAL BUILDING BLOCKS

From now on, we can forget about wires and switches and work with the abstraction of logic blocks operating on 1's and 0's, a simple step that allows us to pass from the realm of engineering into the realm of mathematics. This is the most abstract chapter in the book; it will show you how the methods used to construct a tic-tac-toe machine can be used to construct almost any function. In it, we'll define a powerful set of building blocks: logical functions and finite-state machines. With these elements, it's easy to build a computer.

LOGICAL FUNCTIONS

·····························

In constructing the tic-tac-toe machine, we began by writing the game tree, whch gave us a set of rules for generating the outputs from the inputs. This turns out to be a generally useful method of attack. Once we write down the rules that specify what outputs we want for each combination of inputs, we can build a device that implements these rules using **And**, **Or**, and **Invert** functions. The logic blocks **And**,

Or, and **Invert** form a *universal construction set,* which can be used to implement any set of rules. (These primitive types of logic blocks are sometimes also called *logic gates.*)

This idea of a universal set of blocks is important: it means that the set is general enough to build anything. My favorite toy when I was a child was a set of interlocking plastic bricks called Lego blocks, with which I built all kinds of toys: cars, houses, spaceships, dinosaurs. I loved to play with these blocks, but they were not quite universal, since the only objects you could build with them had a certain squarish, stair-steppy look. Building something with a different shape—a cylinder or a sphere, for example—would require a new type of block. Eventually, I had to switch to another medium in order to build the things I wanted. But the **And**, **Or,** and **Invert** blocks of Boolean logic are a universal construction set for converting inputs to outputs. The best way to see how they form a universal set is to understand a general method for using them to implement rules. To start, we will consider *binary* rules—rules that specify inputs and outputs that are either 1 or 0. The tic-tac-toe machine is a good example of a function specified by binary rules, because the input switches and the output lights are either on or off—that is, either 1 or 0. (Later, we will discuss rules for handling letters, numbers, or even pictures and sounds as inputs and outputs.) Any set of binary rules can be completely specified by showing a table of the outputs for each possible combination of 1's and 0's on the inputs. For example, the rules for the **Or** function are specified by the following table:

	Input A	Input B	Output
	0	0	0
OR Function	0	1	1
	1	0	1
	1	1	1

The **Invert** function is specified by an even simpler table:

	Input	Output
Invert Function	0	1
	1	0

For a binary function with n inputs, there are 2^n possible combinations of input signals. Sometimes we won't bother to specify all of them, because we don't care about certain combinations of inputs. For example, in specifying the function performed by the tic-tac-toe machine, we don't care what happens if the human player plays in all squares simultaneously. This move would be disallowed, and we don't need to specify the function's output for this combination of inputs.

Complex logic blocks are constructed by connecting **And, Or**, and **Invert** blocks. In drawings of the connection pattern, the three blocks are conventionally represented by boxes of different shape (see Figure 9); the lines connecting on the left side represent inputs to the blocks, and the lines connecting on the right represent the output. Figure 10 shows how a pair of two-input **Or** blocks can be connected to form a three-input **Or** function; the output of this function will be 1 if *any one* of its three inputs is 1. It's also possible to string several **And** blocks together in a similar manner to make an **And** block with any number of inputs.

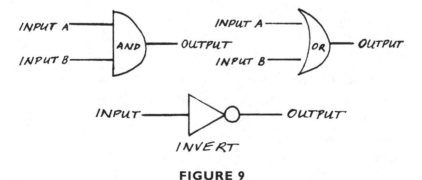

FIGURE 9

And, Or, **and** Invert **Blocks**

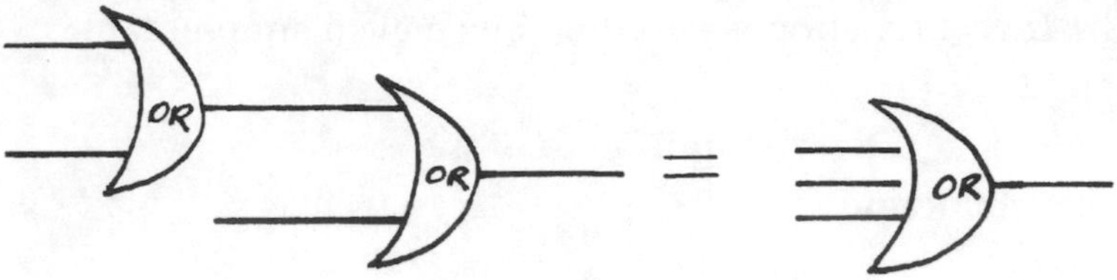

FIGURE 10

A three-input Or block made from a pair of two-input Or blocks

Figure 11 shows how an **And** block can be constructed by connecting an Inverter to the inputs and output of an **Or** block. (Here is De Morgan's theorem again.) The best way to get a feeling for how this works is to trace through the 1's and 0's for every combination of inputs. Notice that this illustration is essentially the same as Figure 6 in the previous chapter. It points up an interesting fact: we don't really need **And** blocks in our universal building set, because we can always construct them out of **Or** blocks and **Inverters**.

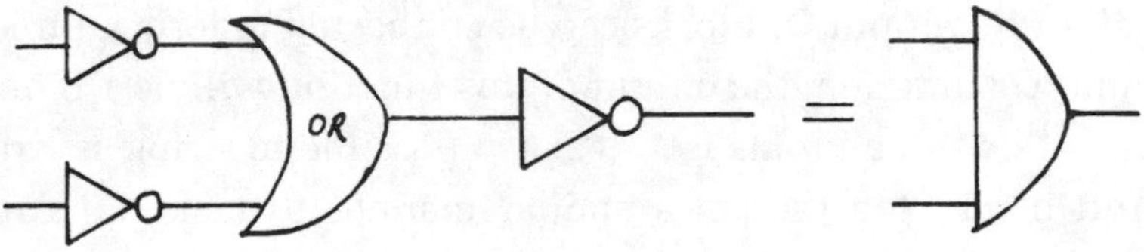

FIGURE 11

Making And **out of** Or

As in the tic-tac-toe playing machine, **And** blocks are used to detect each possible combination of inputs for which the output is 1, while **Or** blocks provide a roster of these combinations. For example, let's start with a simple function of three inputs. Imagine that we want to build a block that allows the three inputs to vote on the output. In this new

block, majority wins—that is, the output will be 1 only if two
or more of the inputs are 1.

	Majority
Inputs	**Output**
A B C	
0 0 0	0
0 0 1	0
0 1 0	0
0 1 1	1
1 0 0	0
1 0 1	1
1 1 0	1
1 1 1	1

Figure 12A shows how this function is implemented.An
And block with the appropriate **Invert** blocks as input is
used to recognize each combination of inputs for which the
output is 1; these blocks are connected by an **Or** block,
which produces the output. This strategy can be used to create any transformation of inputs to outputs:

Of course, this particular method of using a separate **And**
gate to recognize each combination of inputs is not the only
way to implement the function, and it is often not the simplest way. Figure 12B shows a simpler way to produce the
majority function. The great thing about the method described is not that it produces the best implementation but
that it always produces an implementation that works. The
important conclusion to draw is that it is possible to combine **And, Or,** and **Invert** blocks to implement *any* binary
function—that is, any function that can be specified by an
input/output table of 0's and 1's.

Restricting the inputs and output to binary numbers is
not really much of a restriction, because the combinations of
1's and 0's can be used to represent other things—letters,

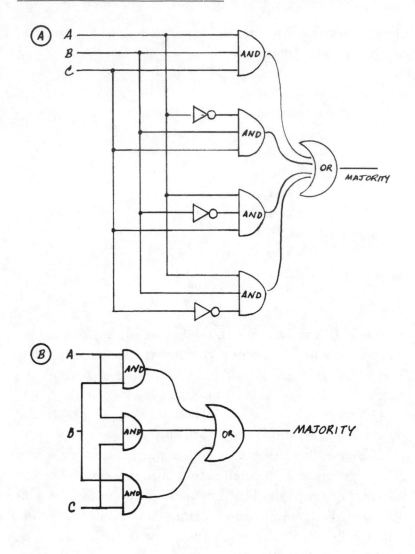

FIGURE 12

How the voting function is implemented by
And, Or, **and** Invert **Blocks**

larger numbers, any entity that can be encoded. As an example of a nonbinary function, suppose we want to build a machine to act as a judge of the children's game of Scissors/Paper/Rock. This is a game for two players in which each chooses, in secret, one of three "weapons"—scissors, paper, or rock. The rules are simple: scissors cuts paper, paper covers rock, rock crushes scissors. If the two children choose the same weapon, they tie. Rather than building a machine that plays the game (which would involve guessing which weapon the opponent is going to choose), we will build a machine that judges who wins. Here's the input/output table for the function that takes the choices as inputs and declares the winner as output. The table encodes the rules of the game:

Input A	Input B	Output
Scissors	Scissors	Tie
Scissors	Paper	A wins
Scissors	Rock	B wins
Paper	Scissors	B wins
Paper	Paper	Tie
Paper	Rock	A wins
Rock	Scissors	A wins
Rock	Paper	B wins
Rock	Rock	Tie

The Scissors-Paper-Rock judging function is a combinational function, but it is not a binary function, since its inputs and output have more than two possible values. To implement this function as a combinational logic block, we must convert it to a function of 1's and 0's. This requires us to establish some convention for representing the inputs and outputs. A simple way to do this would be to use a separate bit for each of the possibilities. There would be three input signals for each weapon: a 1 on the first input represents **Scissors,** a 1 on the second input represents **Rock**, and a 1 on the third input represents **Paper**. Similarly, we could use separate output lines to represent a win for player A, a win for player B, or a tie. So the box would have six inputs and three outputs.

Using three input signals for each weapon is a perfectly good way to build the function, but if we were doing it inside a computer we would probably use some kind of encoding that required a smaller number of inputs and outputs. For example, we could use two bits for each input and use the combination 01 to represent **Scissors**, 10 to represent **Paper**, and 11 to represent **Rock**. We could similarly encode each of the possible outputs using two bits. This encoding would result in the simpler three-input/two-output table shown below:

	A Inputs	B Inputs	Outputs
	01	01	00
	01	10	10
Scissors = 01	01	11	01
Paper = 10	10	01	01
Rock = 11	10	10	00
A wins = 10	10	11	10
B wins = 01	11	01	10
Tie = 00	11	10	01
	11	11	00

Computers can use combinations of bits to represent anything; the number of bits depends on the number of messages that need to be distinguished. Imagine, for example, a computer that works with the letters of the alphabet. Five-bit input signals can represent thirty-two different possibilities ($2^5 = 32$). Functions within the computer that work on letters sometimes use such a code, although they more often use an encoding with seven or eight bits, to allow representation of capitals, punctuation marks, numerals, and so on. Most modern computers use the standard representation of alphabet letters called ASCII (an acronym for American Standard Code for Information Interchange). In ASCII, the sequence 1000001 represents the capital letter **A**, and 1000010 represents the capital **B**, and so on. The convention, of course, is arbitrary.

Most computers have one or more conventions for representing numbers. One of the most common is the *base 2* representation of numbers, in which the bit sequence 0000000 represents zero, the sequence 0000001 represents the number 1, the sequence 0000010 represents 2, and so on. The description of computers as "64-bit" or "32-bit" indicates the number of bit positions in the representation used by the computer's circuits: a 32-bit computer uses a combination of thirty-two bits to represent a base-2 number. The base-2 number system is a common convention, but there is nothing that requires its use. Some computers don't use it at all, and most computers that do also represent numbers in other ways for various purposes. For instance, many computers use a slightly different convention for representing negative numbers and also have a convention called a *floating point* to represent numbers that have decimal points. (The position of the decimal point "floats" relative to the digits, so that a fixed number of digits can be used to represent a wide range of numbers.) The particular representation schemes are often chosen in such a way as to simplify the logic of the circuits that perform arithmetical operations, or to make it easy to convert from one representation to another.

Because any logical function can be implemented as a Boolean logic block, it is possible to build blocks that perform arithmetical operations like addition or multiplication by using numbers with any sort of representation. For instance, imagine that we want to build a functional block that will add numbers on an eight-bit computer. An eight-bit adder block must have sixteen input signals (eight for each of the numbers to be added), and eight output signals for the sum. Since each number is represented by eight bits, there are 256 possible combinations, and each can represent a different number. For example, we could use these combinations to represent the numbers between 0 and 255, or between −100 and +154. Defining the function of the block would be just a matter of writing down the addition table and then converting it to 1's and 0's, using the chosen representation. The table of 1's and 0's could then be converted to **And** and **Or** blocks by the methods described above.

By adding two more inputs to the block, we could use similar techniques to build a block that not only adds but also subtracts, multiplies, and divides. The two extra control inputs would specify which of these operations was to take place. For instance, on every line of the table where the control inputs were 01, we would specify the output to be the sum of the input numbers, whereas in every combination where the control inputs were 10, we would specify the outputs to be the product, and so on. Most computers have logical blocks of this type inside them called *arithmetic units.*

Combining **And**s and **Or**s according to this strategy is one way to build any logical function, but it is not always the most efficient way. Often, by clever design, you can implement a circuit using far fewer building blocks than the preceding strategy requires. It may also be desirable to use other types of building blocks or to design circuits that minimize the delay from input to output. Here are some typical puzzles in logic design: How do you use **And** blocks and **Inverters** to construct **Or** blocks? (Easy.) How do you use a collection of **And** and **Or** blocks, plus only two **Inverters,** to construct the function of three **Inverters**? (Hard, but possible.) Puzzles like this come up in the course of designing a computer, which is part of what makes the process fun.

FINITE-STATE MACHINES
....

The methods I've described can be used to implement any function that stays constant in time, but a more interesting class of functions are those that involve sequences in time. To handle such functions, we use a device called a *finite-state machine.* Finite-state machines can be used to implement time-varying functions—functions that depend not just on the current input but also on the previous history of inputs. Once you learn to recognize a finite-state machine, you'll notice them everywhere—in combination locks, ballpoint pens, even legal contracts. The basic idea of a finite-state machine is to combine a look-up table, constructed using Boolean logic, with a memory device. The memory is used to store a summary of the past, which is the *state* of the finite-state machine.

A combination lock is a simple example of a finite-state machine. The state of a combination lock is a summary of the sequence of numbers dialed into the lock. The lock doesn't remember all the numbers that have ever been dialed into it, but it does remember enough about the most recent numbers to know when they form the sequence that will open the lock. An even simpler example of a finite-state machine is the retractable ballpoint pen. This finite-state machine has two possible states—extended and retracted— and the pen remembers whether its button has been pressed an odd or an even number of times. All finite-state machines have a fixed set of possible states, a set of allowable inputs that change the state (clicking a pen's button, or dialing a number into a combination lock), and a set of possible outputs (retracting or extending the ballpoint, opening the lock). The outputs depend only on the state, which in turn depends only on the history of the sequence of inputs.

Another simple example of a finite-state machine is a counter, such as the tally counter on a turnstile indicating the number of people who have passed through. Each time a new person goes through, the counter's state is advanced by one. The counter is a *finite* state because it can only count up to a certain number of digits. When it reaches its maximum count—say, 999—the next advance will cause it to return to zero. Odometers on automobiles work like this. I once drove an old Checker cab with an odometer that read 70,000, but I never knew if the cab had traveled 70,000 miles, 170,000 miles, or 270,000 miles, because the odometer had only 100,000 states; all those histories were equivalent as far as the odometer was concerned. This is why mathematicians often define a state as "a set of equivalent histories."

Other familiar examples of finite-state machines include traffic lights and elevator-button panels. In these machines, the sequence of states is controlled by some combination of an internal clock and input buttons such as the "Walk" button at the crosswalk and the elevator call and floor-selection buttons. The next state of the machine depends not only on the previous state but also on the signals that come from the input button. The transition from one state to another is determined by a fixed set of rules, which can be summarized by a simple state diagram showing the transition between states. Figure 13 shows a state diagram for a traffic-light

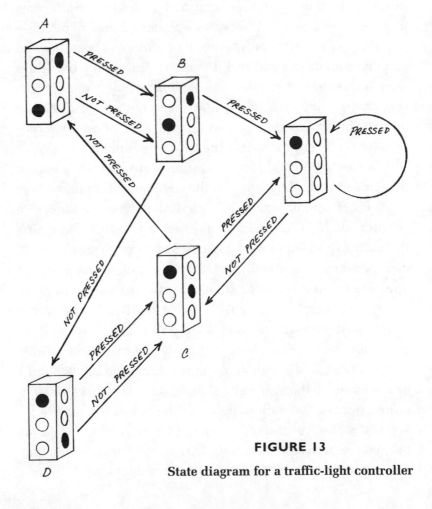

FIGURE 13

State diagram for a traffic-light controller

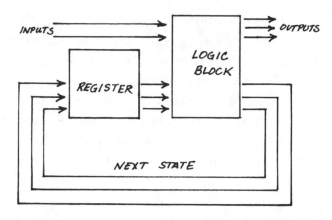

FIGURE 14

Finite-state machine, with logic block feeding register

controller at an intersection where the light turns red in both directions after the Walk button is pressed. Each drawing of light represents a state and each arrow represents a transition between states. The transition depends on whether or not the "walk" button is pressed.

To store the state of the finite-state machine, we need to introduce one last building block—a device called a *register,* which can be used to store bits. An n-bit register has n inputs and n outputs, plus an additional timing input that tells the register when to change state. Storing new information is called "writing" the state of the register. When the timing signal tells the register to write a new state, the register changes its state to match the inputs. The outputs of the register always indicate its current state. Registers can be implemented in many ways, one of which is to use a Boolean logic block to steer the state information around in a circle. This type of register is often used in electronic computers, which is why they lose track of what they're doing if their power is interrupted.

A finite-state machine consists of a Boolean logic block connected to a register, as shown in Figure 14. The finite-state machine advances its state by writing the output of the

Boolean logic block into the register; the logic block then computes the next state, based on the input and the current state. This next state is then written into the register on the next cycle. The process repeats in every cycle.

The function of a finite-state machine can be specified by a table that shows, for every state and every input, the state that follows. For example, we can summarize the operation of the traffic-light controller by the following table:

Inputs:			*Outputs:*	
Walk Button	**Current State**	**Main Road**	**Cross Road**	**Next State**
Not Pressed	A	Red	Green	B
Not Pressed	B	Red	Yellow	D
Not Pressed	C	Yellow	Red	A
Not Pressed	D	Green	Red	C
Not Pressed	Walk	Walk	Walk	D
Pressed	A	Red	Green	B
Pressed	B	Red	Yellow	Walk
Pressed	C	Yellow	Red	Walk
Pressed	D	Green	Red	C
Pressed	Walk	Walk	Walk	Walk

The first step in implementing a finite-state machine is to generate such a table. The second step is to assign a different pattern of bits to each state. The five states of the traffic-light controller will require three bits. (Since each bit doubles the number of possible patterns, it is possible to store up to 2^n states using n bits.) By consistently replacing each word in the preceding table with a binary pattern, we can convert the table to a function that can be implemented with Boolean logic.

In the traffic-light system, a timer controls the writing of the register, which causes the state to change at regular intervals. Another example of a finite-state machine that advances its state at regular intervals is a digital clock. A digital clock with a seconds indicator can be in one of 24 x 60 x 60 = 86,400 possible display states—one for each second of the day. The timing mechanism within the clock causes it to advance its state exactly once per second. Many other types of digital computing devices, including most general-purpose computers, also advance their state at regular intervals, and the rate at which they advance is called the *clock rate* of the machine. Within a computer, time is not a continuous flow but a fixed sequence of transitions between states. The clock rate of the computer determines the rate of these transitions, hence the correspondence between physical and computational time. For instance, the laptop computer on which I am writing this book has a clock rate of 33 megahertz, which means that it advances its state at a rate of 33 million times per second. The computer would be faster if the clock rate were higher, but its speed is limited by the time required for information to propagate through the logic blocks to compute the next state. As technology improves, the logic tends to become faster and the clock rate increases. As I write these words, my computer is state-of-the-art, but by the time you read this book computers with 33 megahertz clock rates will probably be considered slow. This is one of the wonders of silicon technology: as we learn to make computers smaller and smaller, the logic becomes faster and faster.

One reason finite-state machines are so useful is that they can recognize sequences. Consider a combination lock that opens only when it is given the sequence 0–5–2. Such a lock, whether it is mechanical or electronic, is a finite-state machine with the state diagram shown in Figure 15.

A similar machine can be constructed to recognize any finite sequence. Finite-state machines can also be made to recognize sequences that match certain patterns. Figure 16 shows one that recognizes any sequence starting with a 1, followed by a sequence of any number of 0s, followed by a 3. Such a combination will unlock the door with the combination 1–0–3, or a combination such as 1–0–0–0–3, but *not* with the combination 1–0–2–3, which doesn't fit the pattern. A more complex

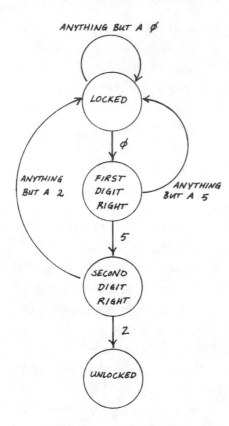

FIGURE 15

State diagram for a lock
with combination 0-5-2

finite-state machine could recognize a more complicated pattern, such as a misspelled word within a stream of text.

As powerful as they are, finite-state machines are not capable of recognizing all types of patterns in a sequence. For instance, it is impossible to build a finite-state machine that will unlock a lock whenever you enter any palindrome—a sequence that is the same forward and backward, like 3–2–1–1–2–3. This is because palindromes can be of any length, and to recognize the second half of a palindrome you need to remember every character in the first half. Since there are infinitely many possible first halves, this would require a machine with an infinite number of states.

A similar argument demonstrates the impossibility of building a finite-state machine that recognizes whether a given English sentence is grammatically correct. Consider the sim-

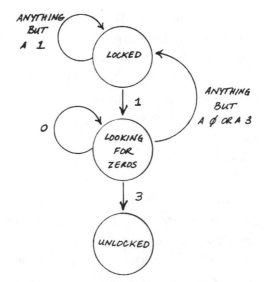

FIGURE 16

State diagram to
recognize sequences
like 1,0,3 and 1,0,0,0,3

ple sentence "Dogs bite." The meaning of this sentence can be
changed by putting a qualifier between the noun and the verb;
for instance, "Dogs that people annoy bite." This sentence can
in turn be modified by putting another phrase in the middle:
"Dogs that people with dogs annoy bite." Although the meaning of such sentences might be expressed more clearly, and
although they become increasingly difficult to understand,
they are grammatically correct. In principle, this process of
nesting phrases inside of one another can go on forever, producing absurd sentences like "Dogs that dogs that dogs that
dogs annoy ate bit bite." Recognizing such a sentence as grammatically correct is impossible for a finite-state machine, and
for exactly the same reason it's difficult for a person: you need
a lot of memory to keep track of all those dogs. The fact that
human beings seem to have trouble with the same kinds of
sentences that stump finite-state machines has caused some
people to speculate that we may have something like a finite-
state machine inside our head for understanding language. As
you will see in the next chapter, there are other types of computing devices that seem to fit even more naturally with the
recursive structure of human grammar.

I was introduced to finite-state machines by my mentor Marvin Minsky. He presented me with the following famous puzzle, called the *firing squad problem*: You are a general in charge of an extremely long line of soldiers in a firing squad. The line is too long for you to shout the order to "fire," and so you must give your order to the first soldier in the line, and ask him to repeat to the next soldier and so on. The hard part is that all the soldiers in the line are supposed to fire at the same time. There is a constant drumbeat in the background; however, you can't even specify that the men should all fire after a certain number of beats, because you don't know how many soldiers are in the line. The problem is to get the entire line to fire simultaneously; you can solve it by issuing a complex set of orders which tells each soldier what to say to the soldiers on either side of him. In this problem, the soldiers are equivalent to a line of finite-state machines with each machine advancing its state by the same clock (the drumbeat), and each receiving input from the output of its immediate neighbors. The problem is therefore to design a line of identical finite-state machines that will produce the "fire" output at the same time in response to a command supplied at one end. (The finite-state machines at either end of the line are allowed to be different from the others.) I won't spoil the puzzle by giving away the solution, but it can be solved using finite-state machines that have only a few states.

Before showing you how Boolean logic and finite-state machines are combined to produce a computer, I'll skip ahead in this bottom-up description and tell you where we're going. The next chapter starts by setting out one of the highest levels of abstraction in the function of a computer, which is also the level at which most programmers interact with the machine.

CHAPTER 3

....

PROGRAMMING

The magic of a computer lies in its ability to become almost anything you can imagine, as long as you can explain exactly what that is. The hitch is in explaining what you want. With the right programming, a computer can become a theater, a musical instrument, a reference book, a chess opponent. No other entity in the world except a human being has such an adaptable, universal nature. Ultimately all these functions are implemented by the Boolean logic blocks and finite-state machines described in the previous chapter, but the human computer programmer rarely thinks about these elements; instead, programmers work with a more convenient tool called a *programming language.*

Just as Boolean logic and finite-state machines are the building blocks of computer hardware, a programming language is a set of building blocks for constructing computer software. Like a human language, a programming language has a vocabulary and a grammar, but unlike a human language there is an exact meaning in the programming language for every word and sentence. Most programming languages are universal, in the same sense that Boolean logic is universal: they can be used to describe anything a computer can do. Anyone who has ever written a program—or debugged a program—knows that telling a computer what you want it to do is not as easy as it sounds. Every detail of the computer's desired operation must be precisely described. For instance, if you tell an accounting program to bill your clients for the amount that each owes, then the computer will send out a weekly bill for $0.00 to clients who owe nothing. If you tell the computer to send a threatening letter to clients who have not paid, then clients who owe nothing will receive threatening letters until they send in payments of $0.00. Avoiding this kind of misunderstanding is what computer programming is all about. The programmer's art is the art of saying exactly what you want. In this example, it means making a distinction between clients who have not sent in any money and clients who actually *owe* money. To paraphrase Mark Twain, the difference between the right program and the almost-right program is like the difference between lightning and a lightning bug—the difference is just a bug.

A skilled programmer is like a poet who can put into words those ideas that others find inexpressible. If you are a poet, you assume a certain amount of shared knowledge and experience on the part of your reader. The knowledge and experience that the programmer and the computer have in common is the meaning of the programming language. How the computer "knows" the meaning of the programming language will be described later; first, we will discuss the grammar, vocabulary, and idioms of such languages.

TALKING TO THE COMPUTER
....

There are many different programming languages. The main reasons for this diversity are history, habit, and taste, but different programming languages also exist because they are good at describing different kinds of things. Each language has its own syntax. You need to learn the syntax in order to write the language, but (like spelling and punctuation in a human language) syntax is not fundamental to the meaning or the expressive power of the language. What is important to the expressive power of the language is the vocabulary— the so-called *primitives* of the language—and the way the primitives can be combined to define new concepts.

Programming languages describe the manipulation of data, and one way in which these languages differ is in the kinds of data they can manipulate. The earliest computer languages were designed primarily to manipulate numbers and sequences of characters. Latter-day programming languages can manipulate words, pictures, sounds, and even other computer programs. But no matter what sort of data the language is designed to handle, it typically provides a way of reading the data's elements into the computer, taking the data apart, putting them together, modifying them, comparing them, and giving them names.

It's probably easier to illustrate these abstractions by describing a particular computer language—I've chosen Logo, which was designed by the educator and mathematician Seymour Papert as a computer language for children. Children can write programs in Logo which create and manipulate pictures, words, numbers, and sounds. Although the language is simple enough to be used by a ten-year-old, it embodies many of the features of the most sophisticated computer languages, including the ability to write programs that manipulate other programs. Logo is also an *extensible* language—that is, you can use Logo to define new words in Logo.

One of the simplest types of programs to write in Logo is a procedure for drawing a picture. You do this by giving directions to an imaginary turtle that lives on the screen. The turtle serves as a pen, moving around on the screen and leaving lines behind it. When the computer starts up, the turtle is found at the center of the screen, facing upward. If the child types the command FORWARD 10, the turtle takes ten steps forward—that is, upward—drawing a line ten units long. The

FIGURE 17

FORWARD 10, RIGHT 45,
FORWARD 10

number 10 following the FORWARD command is called a *parameter;* in this case, the parameter tells the turtle how many steps to take. To draw a line in a different direction, the child must turn the turtle. The command RIGHT 45 will point the turtle 45 degrees (another parameter) to the right from its last heading. The next FORWARD command, with its parameter, will then draw a line in the new direction.

The child can use commands like FORWARD, BACK-WARD, RIGHT, and LEFT to move the turtle around the screen and draw pictures, but this involves a lot of typing and soon becomes tedious. What makes the language interesting is its ability to define new words: for example, here's how the child could teach the turtle (program the computer) to draw a square:

```
TO SQUARE
FORWARD 10
RIGHT 90
FORWARD 10
RIGHT 90
FORWARD 10
RIGHT 90
FORWARD 10
END
```

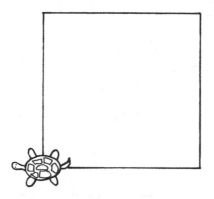

FIGURE 18

A square

Having defined the word "square," the child can then draw a square with ten units on a side simply by typing a new command: SQUARE. (Needless to say, the name "square" is arbitrary; the child could just as well have called the procedure BOX or XYZ and the turtle would do exactly the same thing. When children discover this, they often enjoy "fooling" the computer—for instance, calling the procedure that draws a square TRIANGLE, and vice versa.)

Once the word "square" has been defined, it becomes part of the computer's vocabulary and can then be used to define other words. For example,

    TO WINDOW
    SQUARE
    SQUARE
    SQUARE
    SQUARE
    END

Each square will be drawn in a different place, because the procedure for drawing the square leaves the turtle rotated 90 degrees. In computer terms, SQUARE is a *subroutine* of the

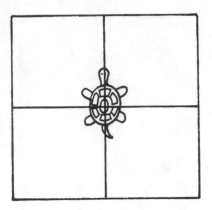

FIGURE 19

A window, made of four squares

program WINDOW, which *calls* it. The subroutine SQUARE, in turn, is defined using the *primitives* FORWARD and RIGHT. User-defined words in Logo can take parameters, too. For instance, a child can specify various sizes of square by specifying what parameter determines the length of each side.

```
TO SQUARE :SIZE
FORWARD :SIZE
RIGHT 90
FORWARD :SIZE
RIGHT 90
FORWARD :SIZE
RIGHT 90
FORWARD :SIZE
END
```

The colon in front of the command SIZE is an example of syntax. In Logo, the colon indicates that the word that follows is the name of a parameter, representing something else—in this case, the number to be supplied each time we "call" the subroutine named SQUARE. When SQUARE is

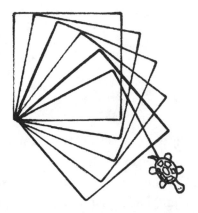

FIGURE 20

A design, drawing in progress

defined this way, the command SQUARE 15 will tell the computer to draw a square fifteen units on a side. The parameter name SIZE is, again, an arbitrary name and has meaning only within the definition of SQUARE: if all five occurrences of SIZE were replaced with X, say, the subroutine would do exactly the same thing.

There are other ways of writing a subroutine to draw a square. For instance, you can instruct the turtle to turn left four times, or to go backward four times. What's interesting is that it doesn't matter how SQUARE is defined; all that matters is what the subroutine draws and where it leaves the turtle. Other programs can call the SQUARE no matter how it is defined and whether or not it is a user-defined word or one of the language's primitives. In extending the language, the programmer uses the power of functional abstraction to create new building blocks.

One of the tricks that Logo-using children discover is that they can insert a word inside its own definition, a practice known as *recursion*. For instance, a child might generate a circular design consisting of many rotated squares, as follows:

```
TO DESIGN
SQUARE
RIGHT 10
DESIGN
END
```

The computer follows the DESIGN command by drawing a square, then turning the turtle 10 degrees and drawing another "design" in the same way. In this case, the recursive definition of "design" has a problem: it goes on forever. Each time the computer responds to the DESIGN command it draws a square, and then it must go on to another design, and on and on ad infinitum—rather as in the story of the guru who claimed that the earth was sitting on the back of (as it happens) a giant turtle. "And what is the turtle sitting on?" asked a student. "Another turtle," replied the guru. "And that turtle?" asked the student, beginning to grow skeptical. "It's no use asking," said the guru. "It's turtles all the way down."

The computer, in drawing the design, goes through the same process that human beings go through in trying to imagine the infinite stack of giant turtles, but the computer is not smart enough to notice that it's not getting anywhere. It will not halt until it is interrupted—an example of a common sort of program behavior called an *infinite loop.* Programmers often create infinite loops accidentally, and (as we shall see) it can be extremely difficult to predict when such loops will occur. This particular infinite loop is easily avoided by writing the program with a parameter specifying how many squares to draw.

```
TO DESIGN :NUMBER
SQUARE
RIGHT 10
IF :NUMBER = 1 STOP ELSE DESIGN :NUMBER −1
END
```

Thus defined, the DESIGN subroutine will do one of two things, depending on whether its parameter is 1 or some higher number. DESIGN 1 will draw just one square, but DESIGN 5, say, will draw a square, rotate, and then draw a DESIGN 4. DESIGN 4 will draw a square and then a DESIGN 3, and so on down to DESIGN 1, which will draw a square and stop.

This kind of recursive definition with a changing parameter is useful for producing anything that has a self-similar structure. A picture that contains a picture of itself is an example of a recursive, self-similar structure; such structures are commonly known as *fractals*. In the real world, self-similar structures don't go on forever: for instance, each branch of a tree looks a lot like a smaller tree, and each of these smaller branches has branches that look like still smaller trees. This recursion goes on for several levels, but eventually the branches are so small that they do not have branches of their own.

A recursive Logo program for drawing a tree is shown below. This may give you some idea of how a computer program can contain an element of poetry—although in this case the theme is somewhat obscured by the details of positioning the turtle and bringing it back to the starting point. Here's a rough translation of what the program says: "A big tree is a stick with two smaller trees on top, but a little tree is just a stick." The picture produced by the tree program is shown next to it.

```
TO TREE :SIZE
FORWARD :SIZE
IF :SIZE <1 STOP ELSE TWO-
TREES SIZE/2
BACK :SIZE
END
```

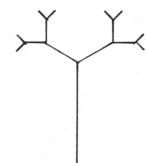

FIGURE 21

A tree

```
TO TWO-TREES :SIZE
LEFT 45
TREE :SIZE
RIGHT 90
TREE :SIZE
LEFT 45
END
```

This technique of defining things recursively turns out to be very powerful. Many of the types of data we like to manipulate—in particular, computer programs themselves—have recursive structures. Recursive definitions are extremely convenient for specifying operations on recursive data. The typical recursive definition has two parts—the first part describes what is to happen in a particular simple case, and the second describes how a more complex case can be reduced to something simpler. In the recursive tree, for example, the simple case is the tree smaller than 1 and the more complex case is the tree composed of a trunk and two small trees.

Another example is the definition of a palindrome, which we can define as follows: A word is a palindrome if it has less than two letters (the simple case) or if its first and last letter are the same and the letters in the middle form a palindrome (the recursive step). The simplest way to write a Logo program for recognizing palindromes would be to use this recursive definition.

There are many other computer languages: LISP, Ada, FORTRAN, C, ALGOL, and the like; most of the names are obscure acronyms (such as FORTRAN, for FORmula TRANslation, and LISP, for LISt Processing). Although these languages differ from Logo in details of vocabulary and syntax, they can all express the same kinds of procedures. Some, like FORTRAN, are limited in their ability to define operations recursively or to manipulate non-numerical data. Others, like C and LISP, allow the programmer to manipulate the underlying bits representing the data, which gives the programmer more power—and more opportunity to make mistakes. In C, for example, it's perfectly possible to multiply two alphabetic characters; the result of this nonsensical operation will depend on the binary representation used by the machine. Languages like LISP offer the abstract as well as the lower-level functions. As a friend of mine, the computer scientist Guy Steele, once put it, "LISP is a high-level language, but you can still feel the bits sliding between your toes."

More recently, a new generation of languages has begun to emerge. These languages—Small-Talk, C++, Java—are *object-oriented.* They treat a data structure—for instance, a picture to be drawn on the screen—as an "object" with its own internal state, such as where it is to be drawn or what color it is. These objects can receive instructions from other objects. To understand why this is useful, imagine that you are writing a program for a video game involving bouncing balls. Each ball on the screen is defined as a different object. The program specifies rules of behavior that tell the object how to draw itself on the screen, move, bounce, and interact with other objects in the game. Each ball will exhibit similar behavior, but each will be in a slightly different state, because each will be in its own position on the screen and will have its own color, velocity, size, and so forth.

The most important advantage of an object-oriented programming language is that the objects—for instance, various objects in a video game—can be specified independently and then combined to create new programs. Writing a new object-oriented program sometimes feels a bit like throwing a bunch of animals into a cage and watching what happens. The behavior of the program *emerges*, as a result of the inter-actions of the programmed objects. For this reason, as well as the fact that object-oriented languages are relatively new, you might think twice about one for writing a safety-critical system that flies an airplane.

Learning a programming language is not nearly as difficult as learning a natural human language. Generally, once you have learned two or three, you can pick up others in a matter of a few hours, since the syntax is relatively simple and the vocabularies are rarely more than a few hundred words. But, as is true of a human language, there's a big difference between being able to understand the language and being able to write it well. Every computer language has its Shakespeares, and it is a joy to read their code. A well-written computer program possesses style, finesse, even humor—and a clarity that rivals the best prose.

MAKING THE CONNECTION

How can finite-state machines be used to carry out instructions written in a language like Logo? To answer this question, we go back to a more detailed level of discussion, involving Boolean logic. There are three major steps in this connection between finite-state machines and Logo: *first,* we will see how a finite-state machine can be extended, by adding a storage device called a *memory,* which will allow the machine to store the definitions of what it's asked to do; *second,* we will see how this extended machine can follow instructions written in *machine language,* a simple language that specifies the machine's operations; and *third,* we will see how machine language can instruct the machine to interpret the programming language—for instance, Logo. The rest of the chapter describes how all this works in some detail—far more detail than is strictly necessary to understand the rest of the book. The reader should not feel compelled to understand every step. The important thing to appreciate is how the layers of functional abstraction build upon one another, as is summarized in the last paragraph of the chapter.

A computer is just a special type of finite-state machine connected to a memory. The computer's memory—in effect, an array of cubbyholes for storing data—is built of *registers,* like the registers that hold the states of finite-state machines. Each register holds a pattern of bits called a *word*, which can be read (or written) by the finite-state machine. The number of bits in a word varies from computer to computer, but in a modern microprocessor (as I write this) it is usually eight, sixteen, or thirty-two bits. (Word sizes will probably grow with improvement in technology.) A typical memory will have millions or even billions of these registers, each holding a single word. Only one of the registers in the memory is accessed at a time—that is, only the data in one of the memory registers will be read or written on each cycle of the finite-state machine. Each register in the memory has a different *address*—a pattern of bits by means of which you can access it—so registers are referred to as *locations in memory*. The memory contains Boolean logic blocks, which decode the address and select the location for reading or writing. If data are to be written at this memory location, these logic blocks store the new data into the addressed register. If the register is to be read, the logic blocks steer the data from the addressed register to the memory's output, which is connected to the input of the finite-state machine.

Some of the words stored in the memory represent data to be operated upon, like numbers and letters. Others represent *instructions* that tell the machine what sequence of operations to perform. The instructions are stored in machine language, which, as noted, is much simpler than a typical programming language. Machine language is interpreted directly by the finite-state machine. In the type of computer we will describe, each instruction in machine language is stored in a single word of memory, and a sequence of instructions is stored in a block of sequentially numbered memory locations. These sequences of machine-language instructions are the simplest kind of software within the computer.

The finite-state machine repeatedly executes the following sequence of operations: (1) *read* an instruction from the memory, (2) *execute* the operation specified by that instruc-

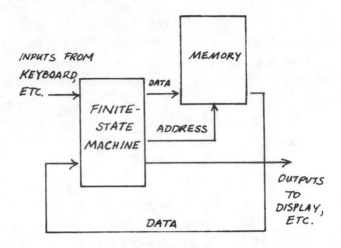

FIGURE 22

Finite-state machine connected to a memory

tion, and (3) *calculate* the address of the next instruction. The sequence of states necessary to do this is built into the Boolean logic of the machine, and the instructions themselves are specific patterns of bits—patterns that cause the finite-state machine to perform various operations on the data in the memory. For instance, the **Add** instruction is a unique pattern of bits that specifies which two registers in the memory are to be added together. Upon recognizing this pattern, the finite-state machine will go through a sequence of states that cause it to read the memory locations to be summed, add the numbers together, and write the sum back to the memory.

There are two basic types of instructions in most computers: processing instructions and control instructions. The *processing instructions* move data to and from the memory and combine them to perform arithmetic and logical functions. The addresses of the memory locations, or registers, are specified by the processing instructions. Typically, these instructions refer to only a few registers directly; other registers are referenced indirectly, because their addresses are stored in other registers. For example, a **Move** instruction might move the data in register 1 to the address specified in register 2. If register 2 holds a pattern of bits that specifies the number 1,234, then the data will be moved to register 1,234. Other processing instructions combine data among the memory registers. There are also instructions to perform Boolean functions—**And**, **Or**, or **Invert**—on the patterns of bits in the registers.

The *control instructions* determine the address of the next instruction to be fetched; this address is stored in a special register called the *program counter*. Normally, instructions are fetched sequentially from successive memory locations, so the address in the program counter increases by 1 after each successive instruction. Control instructions allow some other number to be loaded into the program counter, thereby affecting the sequence to be executed. The simplest control instruction is the **Jump** instruction, which stores a specified address in the program counter so that the next instruction will be fetched from that address. A variation of the **Jump** instruction is a conditional **Jump**, which loads the program counter with a different address only when a specified condition has been met—such as the patterns in two registers being the same. If the condition is not met, then the conditional **Jump** will have no effect and the next instruction will be fetched sequentially.

If the same sequence of instructions needs to be executed over and over repeatedly, then a conditional **Jump** can be used at the end of the sequence to move the program counter back to the beginning as many times as is necessary. This operation is called a *loop,* and we saw an example of it in the description of programming in Logo. The execution of the sequence will be repeated until the condition on which the jump depends is no longer satisfied. If a set of instructions is to be repeated ten times, say, one of the memory registers can be used to count the number of iterations of the loop.

Exactly which instructions a computer recognizes varies from computer to computer. Computer designers can (and do) spend years arguing about what makes an optimal instruction set. A typical argument is over the comparative merits of a *reduced instruction set computer* (RISC), which uses a simple, minimal set of instructions, and a *complex instruction set computer* (CISC), which employs a rich, complex, and powerful set of instructions. This argument has little import to the programmer, however, since any reasonable instruction set can simulate any other. Historically, the commercial success of one or another sort of computer seems to have had almost nothing to do with the complexity of the instruction set or any other detail of internal design. In fact, some of the most successful computers—such as the microprocessors used in most personal computers—are generally regarded by computer designers as having poorly designed instruction sets. The details of machine design are of little importance to the users of the machines.

One reason the complexity of the instruction set doesn't much matter has to do with the subroutines. Subroutines allow sequences of instructions to be used over and over from many places within the program. In effect, the subroutine-calling convention allows the programmer to define new instructions by using sequences of other instructions. The program accesses a subroutine by using a **Jump** instruction to load the program counter with the address of the subroutine; but before doing so, the computer saves the previous contents of the program counter in a special memory location. At the end of the subroutine, another instruction reads this return address and jumps back to the location from which the subroutine was called.

This process of subroutine calling can work recursively, in the sense that subroutine sequences may have jumps to other subroutines within them, and so on. A subroutine can even call itself, in a recursively defined function. In order to keep track of this nesting of subroutines, the computer needs a systematic way of storing the return addresses, so that it will know where to return when each subroutine is completed. It cannot just store all the return addresses in the same special location, because when subroutines are nested, it needs to remember more than one return address. Normally, the computer stores return addresses in a group of sequential locations known as a *stack*. The most recent return address is stored at the "top of the stack." The memory stack works just like a stack of dinner plates: items are always added or removed from the top—a last-in, first-out storage system that works perfectly for storing the return addresses of nested subroutines, because a subroutine is never finished until all of its nested subroutines are finished.

Some subroutines are so useful that they are always loaded into the computer. This set of subroutines is called the *operating system*. Useful operating-system subroutines include those that write or read characters typed on the keyboard, or that draw lines on the screen, or otherwise interact with the user. The computer's operating system determines most of the look and feel of the interface to the user. It also governs the interface between the computer and whatever program is being run, since the operating system's subroutines provide the program with a set of operations that are richer and more complex than the machine-language instructions.

In fact, as long as the same pattern of bits produces the same effect, the programmer doesn't care whether a function is implemented by the computer hardware or by the operating-system software. The same program operating on two different types of computers might in one case perform an arithmetical operation in the hardware and in the other perform it by means of an operating-system subroutine. Similarly, the operating system of one type of computer may allow it to emulate the entire instruction set of another type of computer. Computer manufacturers sometime use such emulation to make the newer model of their computers act like the earlier models, so that older software can run without modification.

The operating system normally includes all the subroutines that perform input/output—that is, operations allowing a program to interact with the outside world. This interaction is effected by connecting certain locations in the computer's memory to input devices such as a keyboard or a mouse and output devices such as a video display terminal. For example, the space bar on the keyboard might be wired to memory register 23, so that the data read from address 23 is 1 if the space bar is pressed and 0 if it is not. Another memory register might control the color displayed on a certain dot on the screen. If every dot on the screen displays data stored in a different memory location, the computer can draw any pattern on the screen just by writing the appropriate pattern into memory.

With the exception of its input/output mechanisms, the computer we've just described is simply a finite-state machine connected to a memory. Both these elements can be constructed entirely from registers and Boolean logic blocks, using the techniques described in chapters 1 and 2. The finite-state machine that controls the computer is complicated, but it is no different in principle from the finite-state machine that controls a traffic light. Designing the machine is simply a matter of going through the details of memory data, address, and state sequences for each instruction to be executed, and then converting this state table into Boolean logic. Recall that since both the finite-state machine and the memory are made of registers and blocks of logic, both can be implemented by a number of technologies: electronics, hydraulics, sliding sticks.

TRANSLATING THE LANGUAGE
·······································

So we have established a chain of connections between the technology and the instructions. But how do the instructions execute a program written in a language—Logo, for instance—when that language is written in words and the instructions are patterns of bits? The answer is that the necessary translation is performed by the computer itself.

The translation process performed by the computer is similar to the process that a patient, meticulous human translator would use to translate a document written in an unfamiliar language, given a dictionary written in the language itself. The human translator can look up the meaning of any unknown word in the dictionary, and if words in the dictionary definition are also unknown, these words can be looked up as well. This process continues until the translator reaches a definition couched in words whose meanings are known. In this analogy, the translator's (that is, the computer's) dictionary is the program, and the words known to the computer are the aforementioned primitives of the program language. These primitives are defined directly, as simple sequences of machine instructions. For instance, when the computer looks up the definition of the Logo language primitive FORWARD, it finds the sequence of machine instructions that will draw the appropriate line on the screen.

To understand how a computer translates Logo primitives into machine language, it is helpful to understand the conventions that a computer uses to represent the Logo programs within its memory. One way to store a Logo program in a computer's memory is as a sequence of characters in adjacent memory locations, with each character being stored at a single location. In its memory, the computer keeps a directory of the addresses of the instruction sequences corresponding to each command name. This directory is stored in memory as a list of names paired with their addresses. The computer is able to find the location of an object with a given name by searching the directory for the name and finding the corresponding address. When the computer is asked to execute a particular command, it looks up the name in the directory to find out where its definition is stored.

Some of this process of looking things up and finding the corresponding sequences of machine language can be done before the program is executed. This saves time, because if the program is going to be executed more than once, there's no point in looking up the same things over and over again. When most of the work of conversion is done beforehand, the translation process is called *compilation,* and the program that performs the compilation is called a *compiler.* If most of the work is done while the program is being executed, then the process is called interpretation, and the program is called an *interpreter.* There is no hard and fast line between the two.

WELCOME TO THE HIERARCHY

We are now in a position to summarize how a computer works, from top to bottom. Most readers will have lost track of the details, but *remember that it is not important to remember how every step works!* The important thing to remember is the hierarchy of functional abstractions.

The work performed by the computer is specified by a *program,* which is written in a *programming language.* This language is converted to sequences of *machine-language* instructions by *interpreters* or *compilers,* via a predefined set of subroutines called the *operating system.* The instructions, which are stored in the *memory* of the computer, define the operations to be performed on data, which are also stored in the computer's memory. A *finite-state machine* fetches and executes these instructions. The instructions as well as the data are represented by patterns of *bits.* Both the finite-state machine and the memory are built of storage *registers* and *Boolean logic blocks,* and the latter are based on simple *logical functions,* such as *And, Or,* and *Invert.* These logical functions are implemented by *switches,* which are set up either *in series* or *in parallel*, and these switches control a physical substance, such as water or electricity, which is used to send one of two possible signals from one switch to another: *1* or *0*. This is the hierarchy of abstraction that makes computers work.

CHAPTER 4

···

HOW UNIVERSAL ARE
TURING MACHINES?

What are the limits to what a computer can do? Must all computers be composed of Boolean logic and registers, or might there be other kinds, even more powerful? These questions take us to the most philosophically interesting topics in this book: Turing machines, computability, chaotic systems, Goedel's incompleteness theorem, and quantum computing—topics at the center of most discussions about what computers can and cannot do.

Because computers can do some things that seem very much like human thinking, people often worry that they threaten our unique position as rational beings, and there are some who seek reassurance in mathematical proofs of the limits of computers. There have been analogous controversies in human history. It was once considered important that the Earth be at the center of the universe, and our imagined position at the center was emblematic of our worth. The discovery that we occupied no central position—that our planet was just one of a number of planets in orbit around the Sun—was deeply disturbing to many people at the time, and the philosophical implications of astronomy became a topic of heated debate. A similar controversy arose over evolutionary theory, which also appeared as a threat to humankind's uniqueness. At the root of these earlier philosophical crises was a misplaced judgment of the source of human worth. I am convinced that most of the current philosophical discussions about the limits of computers are based on a similar misjudgment.

TURING MACHINES

The central idea in the theory of computation is that of a *universal computer*—that is, a computer powerful enough to simulate any other computing device. The general-purpose computer described in the preceding chapters is an example of a universal computer; in fact most computers we encounter in everyday life are universal computers. With the right software and enough time and memory, any universal computer can simulate any other type of computer, or (as far as we know) any other device at all that processes information.

One consequence of this principle of universality is that the only important difference in power between two computers is their speed and the size of their memory. Computers may differ in the kinds of input and output devices connected to them, but these so-called peripherals are not essential characteristics of a computer, any more than its size or its cost or the color of its case. In terms of what they are able to do, all computers (and all other types of universal computing devices) are fundamentally identical.

The idea of a universal computer was recognized and described in 1937 by the British mathematician Alan Turing. Turing, like so many other computing pioneers, was interested in the problem of making a machine that could think, and he invented a scheme for a general-purpose computing machine. Turing referred to his imaginary construct as a "universal machine," since at that time the word "computer" still meant "a person who performs computations."

To picture a Turing machine, imagine a mathematician performing calculations on a scroll of paper. Imagine further that the scroll is infinitely long, so that we don't need to worry about running out of places to write things down. The mathematician will be able to solve any solvable computational problem no matter how many operations are involved, although it may take him an inordinate amount of time. Turing showed that any calculation that can be performed by a smart mathematician can also be performed by a stupid but meticulous clerk who follows a simple set of rules for reading and writing the information on the scroll. In fact, he showed that the human clerk can be replaced by a finite-state machine. The finite-state machine looks at only one symbol on the scroll at a time, so the scroll is best thought of as a narrow paper tape, with a single symbol on each line.

Today, we call the combination of a finite-state machine with an infinitely long tape a *Turing machine*. The tape of a Turing machine is analogous to, and serves much the same function as, the memory of a modern computer. All that the finite-state machine does is read or write a symbol on the tape and move back and forth according to a fixed and simple set of rules. Turing showed that any computable problem could be solved by writing symbols on the tape of a Turing machine—symbols that would specify not just the problem but also the method of solving it. The Turing machine computes the answer by moving back and forth across the tape, reading and writing symbols, until the solution is written on the tape.

I find Turing's particular construction difficult to think about. To me, the conventional computer, which has a memory instead of a tape, is a more comprehensible example of a universal machine. For instance, it is easier for me to see how a conventional computer can be programmed to simulate a Turing machine than vice versa. What is amazing to me is not so much Turing's imaginary construct but his hypothesis that there is only one type of universal computing machine. As far as we know, no device built in the physical universe can have any more computational power than a Turing machine. To put it more precisely, any computation that can be performed by any physical computing device can be performed by any universal computer, as long as the latter has sufficient time and memory. This is a remarkable statement, suggesting as it does that a universal computer with the proper programming should be able to simulate the function of a human brain.

LEVELS OF POWER

How can Turing's hypothesis be true? Surely some other kind of computer could be more powerful than the ones we have described. For one thing, the computers we have discussed so far have been binary, that is, they represent everything in terms of 1 and 0. Wouldn't a computer be more powerful if it could represent things in terms of a three-state logic, like **Yes**, **No**, and **Maybe**? No, it would not. We know that a three-state computer would be able to do no more than a two-state computer, because you can simulate the one using the other. With a two-state computer, you can duplicate any operation that can be performed on a three-state computer, by encoding each of the three states as a pair of bits—00 for **Yes**, say, and 11 for **No**, and 01 for **Maybe**. For every possible function in three-state logic, there is a corresponding function in two-state logic which operates on this representation. This is not to say that three-state computers might not have some practical advantage over two-state computers: for instance, they might use fewer wires and therefore might be smaller, or cheaper to produce. But we can say for certain that they would not be able to do anything new. They would just be one more version of a universal machine.

A similar argument holds for four-state computers, or five-state computers, or computers with any finite number of states. But what about computers that compute with analog signals—that is, signals with an infinite number of possible values? For example, imagine a computer that uses a continuous range of voltages to indicate numbers. Instead of just two or three or five possible messages, each signal could carry an infinite number of possible messages, corresponding to the continuous range of voltages. For instance, an analog computer might represent a number between 0 and 1 by a voltage between zero and one volt. The fraction could be represented to any level of precision, no matter the number of decimal places, by using the exact corresponding voltage.

Computers that represent quantities by such analog signals do exist, and in fact the earliest computers worked this way. They are called *analog computers*, to distinguish them from the digital computers we have been discussing, which have a discrete number of possible messages in each signal. One might suppose that analog computers would be more powerful, since they can represent a continuum of values, whereas digital computers can represent data only as discrete numbers. However, this apparent advantage disappears if we take a closer look. A true continuum is unrealizable in the physical world.

The problem with analog computers is that their signals can achieve only a limited degree of accuracy. Any type of analog signal—electrical, mechanical, chemical—will contain a certain amount of noise; that is, at a certain level of resolution, the signal will be essentially random. Any analog signal is bound to be affected by numerous irrelevant and unknown sources of noise: for example, an electrical signal can be disturbed by the random motion of molecules inside a wire, or by the magnetic field created when a light is turned on in the next room. In a very good electrical circuit, this noise can be made very small—say, a millionth the size of the signal itself—but it will always exist. While there are an infinite number of possible signal levels, only a finite number of levels represent meaningful distinctions—that is, represent information. If one part in a million in a signal is noise, then there are only about a million meaningful distinctions in the signal; therefore, information in the signal can be represented by a digital signal that uses twenty bits (2^{20} = 1,048,578). Doubling the number of meaningful distinctions in an analog computer would require making everything twice as accurate, whereas in a digital computer you could double the number of meaningful distinctions by adding a single bit. The very best analog computers have fewer than thirty bits of accuracy. Since digital computers often represent numbers using thirty-two or sixty-four bits, they can in practice generate a much larger number of meaningful distinctions than analog computers can.

Some people might argue that while the noise of an analog computer may not be meaningful, it is not necessarily useless. One can certainly imagine computations that are helped by the presence of noise. Later, for example, we will describe computations requiring random numbers. But a digital computer, too, can generate random noise if randomness is called for in a computation.

RANDOM NUMBERS

How can a digital computer generate randomness? Can a deterministic system like a computer produce a truly random sequence of numbers? In a formal sense, the answer is No, since everything a digital computer does is determined by its design and its inputs. But the same could be said of a roulette wheel—after all, the ball's final landing place is determined by the physics of the ball (its mass, its velocity) and the spinning wheel. If we knew the exact design of the apparatus and the exact "inputs" governing the spin of the wheel and the throw of the ball, we could predict the number on which the ball would land. The outcome appears random because it exhibits no obvious pattern and is difficult, in practice, to predict.

Like the roulette wheel, a computer can produce a sequence of numbers that is random in the same sense. In fact, using a mathematical model, the computer could simulate the physics of the roulette wheel and throw a simulated ball at a slightly different angle each time in order to produce each number in the sequence. Even if the angles at which the computer throws the simulated ball follow a consistent pattern, the simulated dynamics of the wheel would transform these tiny differences into what amounts to an unpredictable sequence of numbers. Such a sequence of numbers is called a *pseudorandom* sequence, because it only appears random to an observer who does not know how it was computed. The sequence produced by a pseudorandom number generator can pass all normal statistical tests of randomness.

A roulette wheel is an example of what physicists call a *chaotic system*—a system in which a small change in the initial conditions (the throw, the mass of the ball, the diameter of the wheel, and so forth) can produce a large change in the state to which the system evolves (the resulting number). This notion of a chaotic system helps explain how a deterministic set of interactions can produce unpredictable results. In a computer, there are simpler ways to produce a pseudorandom sequence than simulating a roulette wheel, but they are all conceptually similar to this model.

Digital computers are predictable and unpredictable in exactly the same senses as the rest of the physical world. They follow deterministic laws, but these laws have complicated consequences that are extremely difficult to predict. It is often impractical to guess what computers are going to do before they do it. As is true of physical systems, it does not take much to make a computation complex. In computers, chaotic systems—systems whose outcomes depend sensitively on the initial conditions—are the norm.

COMPUTABILITY

••••••••••••••••••••

While a universal computer can compute anything that can be computed by any other computing device, there are some things that are just impossible to compute. Of course, it is not possible to compute answers to vaguely defined questions, like "What is the meaning of life?" or questions for which we lack data, like "What is the winning number in tomorrow's lottery?" But there are also flawlessly defined computational problems that are impossible to solve. Such problems are called *noncomputable.*

I should warn you that noncomputable problems hardly ever come up in practice. In fact, it is difficult to find examples of a well-defined noncomputable problem that anybody wants to compute. A rare example of a well-defined, useful, but noncomputable problem is the *halting problem.* Imagine that I want to write a computer program that will examine another computer program and determine whether or not that program will eventually stop. If the program being examined has no loops or recursive subroutine calls, it is bound to finish eventually, but if it does have such constructs the program may well go on forever. It turns out that there is no algorithm for examining a program and determining whether or not it is fatally infected with an endless loop. Moreover, it's not that no one has yet discovered such an algorithm; rather, no such algorithm is possible. The halting problem is noncomputable.

To understand why, imagine for a moment that I do have such a program, called Test-for-Halt, and that it takes the program to be tested as an input. (Treating a program as data may seem strange, but it's perfectly possible, because a program, just like anything else, can be represented as bits.) I could insert the Test-for-Halt program as a subroutine in another program, called Paradox, which will perform Test-for-Halt on Paradox itself. Imagine that I have written the Paradox program so that whatever Test-for-Halt determines, Paradox will do the opposite. If Test-for-Halt determines that Paradox is eventually going to halt, then Paradox is programmed to go into an infinite loop. If Test-for-Halt determines that Paradox is going to go on forever, then Paradox is programmed to halt. Since Paradox contradicts Test-for-Halt, Test-for-Halt doesn't work on Paradox; therefore, it doesn't work on all programs. And therefore a program that computes the halting function cannot exist.

The halting problem, which was dreamed up by Alan Turing, is chiefly important as an example of a noncomputable problem, and most noncomputable problems that do come up in practice are similar to or equivalent to it. But a computer's inability to solve the halting problem is not a weakness of the computer, because the halting problem is inherently unsolvable. There is no machine that can be constructed that can solve the halting problem. And as far as we know, there is nothing that can perform any other computation that cannot be performed by a universal machine. The class of problems that are computable by a digital computer apparently includes every problem that is computable by any kind of device. (This last statement is sometimes called the Church thesis, after one of Turing's contemporaries, Alonzo Church. Mathematicians had been thinking about computation and logic for centuries but—in one of the more dazzling examples of synchrony in science—Turing, Church, and another British mathematician named Emil Post all independently invented the idea of universal computation at roughly the same time. They had very different ways of describing it, but they all published their results in 1937, setting the stage for the computer revolution soon to follow.)

Another noncomputable function, closely related to the halting problem, is the problem of deciding whether any given mathematical statement is true or false. There is no algorithm that can solve this problem, either—a conclusion of Goedel's incompleteness theorem, which was proved by

Kurt Goedel in 1931, just before Turing described the halting problem. Goedel's theorem came as a shock to many mathematicians, who until then had generally assumed that *any* mathematical statement could be proved true or false. Goedel's theorem states that within any self-consistent mathematical system powerful enough to express arithmetic, there exist statements that can neither be proved true nor false. Mathematicians saw their job as proving or disproving statements, and Goedel's theorem proved that their "job" was in certain instances impossible.

Some mathematicians and philosophers have ascribed almost mystical properties to Goedel's incompleteness theorem. A few believe that the theorem proves that human intuition somehow surpasses the power of a computer—that human beings may be able to "intuit" truths that are impossible for machines to prove or disprove. This is an emotionally appealing argument, and it is sometimes seized upon by philosophers who don't like being compared to computers. But the argument is fallacious. Whether or not people can successfully make intuitive leaps that cannot be made by computers, Goedel's incompleteness theorem provides no reason to believe that there are mathematical statements that can be proved by a mathematician but can't be proved by a computer. As far as we know, any theorem that can be proved by a human being can also be proved by a computer. Humans cannot compute noncomputable problems any more than computers can.

Although one is hard pressed to come up with specific examples of noncomputable problems, one can easily prove that most of the possible mathematical functions are noncomputable. This is because any program can be specified in a finite number of bits, whereas specifying a function usually requires an infinite number of bits, so there are a lot more functions than programs. Consider the kind of mathematical function that converts one number into another—the cosine, say, or the logarithm. Mathematicians can define all kinds of bizarre functions of this type: for example the function that converts every decimal number into the sum of its digits. As far as I know, this function is a useless one, but a mathematician would regard it as a legitimate function simply because it converts every number into exactly one other number. It can be proved mathematically that there are infinitely more functions than programs. Therefore, for most functions there is no corresponding program that can compute them. The actual counting involves all kinds of difficulties (including counting infinite things and distinguishing between various degrees of infinity!), but the conclusion is correct: statistically speaking, most mathematical functions are noncomputable. Fortunately, almost all these noncomputable functions are useless, and virtually all the functions we might want to compute are computable.

QUANTUM COMPUTING

As noted earlier, the pseudorandom number sequences produced by computers look random, but there is an underlying algorithm that generates them. If you know how a sequence is generated, it is necessarily predictable and not random. If ever we needed an inherently unpredictable random-number sequence, we would have to augment our universal machine with a nondeterministic device for generating randomness.

One might imagine such a randomness-generating device as being a kind of electronic roulette wheel, but, as we have seen, such a device is not truly random because of the laws of physics. The only way we know how to achieve genuinely unpredictable effects is to rely on quantum mechanics. Unlike the classical physics of the roulette wheel, in which effects are determined by causes, quantum mechanics produces effects that are purely probabilistic. There is no way of predicting, for example, when a given uranium atom will decay into lead. Therefore one could use a Geiger counter to generate truly random data sequences—something impossible in principle for a universal computer to do.

The laws of quantum mechanics raise a number of questions about universal computers that no one has yet answered. At first glance, it would seem that quantum mechanics fits nicely with digital computers, since the word "quantum" conveys essentially the same notion as the word "digital." Like digital phenomena, quantum phenomena exist only in discrete states. From the quantum point of view, the (apparently) continuous, analog nature of the physical world—the flow of electricity, for example—is an illusion caused by our seeing things on a large scale rather than an atomic scale. The good news of quantum mechanics is that at the atomic scale everything is discrete, everything is digital. An electric charge contains a certain number of electrons, and there is no such thing as half an electron. The bad news is that the rules governing how objects interact at this scale are counterintuitive.

For instance, our commonsense notions tell us that one thing cannot be in two places at the same time. In the quantum mechanical world this is not exactly true, because in quantum mechanics nothing can be exactly in any place at all. A single subatomic particle exists everywhere at once, and we are just more likely to observe such a particle at one place than at another. For most purposes, we can think of a particle as being where we observe it to be, but to explain all observed effects we have to acknowledge that the particle is in more than one place. Almost everyone, including many physicists, find this concept difficult to comprehend.

Might we take advantage of quantum effects to build a more powerful type of computer? As of now, this question remains unanswered, but there are suggestions that such a thing is possible. Atoms seem able to compute certain problems easily, such as how they stick together—problems that are very difficult to compute on a conventional computer. For instance, when two hydrogen atoms bind to an oxygen atom to form a water molecule, these atoms somehow "compute" that the angle between the two bonds should be 107 degrees. It is possible to approximately calculate this angle from quantum mechanical principles using a digital computer, but it takes a long time, and the more accurate the calculation the longer it takes. Yet every molecule in a glass of water is able to perform this calculation almost instantly. How can a single molecule be so much faster than a digital computer?

The reason it takes the computer so long to calculate this quantum mechanical problem is that the computer would have to take into account an infinite number of possible configurations of the water molecule to produce an exact answer. The calculation must allow for the fact that the atoms comprising the molecule can be in all configurations at once. This is why the computer can only approximate the answer in a finite amount of time. One way of explaining how the water molecule can make the same calculation is to imagine it trying out every possible configuration simultaneously—in other words, using parallel processing. Could we harness this simultaneous computing capability of quantum mechanical objects to produce a more powerful computer? Nobody knows for sure.

Recently there have been some intriguing hints that we may be able to build a quantum computer that takes advantage of a phenomenon known as *entanglement*. In a quantum mechanical system, when two particles interact, their fates can become linked in a way utterly unlike anything we see in the classical physical world: when we measure some characteristic of one of them, it affects what we measure in the other, even if the particles are physically separated. Einstein called this effect, which involves no time delay, "spooky action at a distance," and he was famously unhappy with the notion that the world could work that way.

A quantum computer would take advantage of entanglement: a one-bit quantum mechanical memory register would store not just a 1 or a 0; it would store a superposition of many 1's and many 0's. This is analagous to an atom being in many places at once: a bit that it is in many states (1 or 0) at once. This is different from being in an intermediate state between a 1 and a 0, because each of the superposed 1's and 0's can be entangled with other bits within the quantum computer. When two such quantum bits are combined in a quantum logic block, each of their superposed states can interact in different ways, producing an even richer set of entanglements. The amount of computation that can be accomplished by a single quantum logic block is very large, perhaps even infinite.

The theory behind quantum computing is well established, but there are still problems in putting it to use. For one thing, how can we use all this computation to compute anything useful? The physicist Peter Shor recently discovered a way to use these quantum effects—at least, in principle—to do certain important and difficult calculations like factoring large numbers, and his work has renewed interest in quantum computers. But many difficulties are still there. One problem is that the bits in a quantum computer must remain entangled in order for the computation to work, but the smallest of disturbances—a passing cosmic ray, say, or possibly even the inherent noisiness of the vacuum itself—can destroy the entanglement. (Yes, in quantum mechanics even a vacuum does strange things.) This loss of entanglement, called *decoherence*, could turn out to be the Achilles heel of quantum mechanical computers. Moreover, Shor's methods seem to work only on a specific class of computations which can take advantage of a fast operation called a generalized Fourier transform. The problems that fit into this category may well turn out to be easy to compute on a classical Turing machine; if so, Shor's quantum ideas would be equivalent to some program on a conventional computer.

If it does become possible for quantum computers to search an infinite number of possibilities at once, then they would be qualitatively, fundamentally more powerful than conventional computing machines. Most scientists would be surprised if quantum mechanics succeeds in providing a kind of computer more powerful than a Turing machine, but science makes progress through a series of surprises. If you're hoping to be surprised by a new sort of computer, quantum mechanics is a good area to keep an eye on.

This leads us back to the philosophical issues touched on at the beginning of the chapter—that is, the relationship between the computer and the human brain. It is certainly conceivable, as at least one well-known physicist has speculated (to hoots from most of his colleagues), that the human brain takes advantage of quantum mechanical effects. Yet there is no evidence whatsoever that this is the case. Certainly, the physics of a neuron depends on quantum mechanics, just as the physics of a transistor does, but there is no evidence that neural processing takes place at the quantum mechanical level as opposed to the classical level; that is, there is no evidence that quantum mechanics is necessary to explain human thought. As far as we know, all the relevant computational properties of a neuron can be simulated on a conventional computer. If this is indeed the case, then it is also possible to simulate a network of tens of billions of such neurons, which means, in turn, that the brain can be simulated on a universal machine. Even if it turns out that the brain takes advantage of quantum computation, we will probably learn how to build devices that take advantage of the same effects—in which case it will still be possible to simulate the human brain with a machine.

The theoretical limitations of computers provide no useful dividing line between human beings and machines. As far as we know, the brain is a kind of computer, and thought is just a complex computation. Perhaps this conclusion sounds harsh to you, but in my view it takes nothing away from the wonder or value of human thought. The statement that thought is a complex computation is like the statement sometimes made by biologists that life is a complex chemical reaction: both statements are true, and yet they still may be seen as incomplete. They identify the correct components, but they ignore the mystery. To me, life and thought are both made all the more wonderful by the realization that they emerge from simple, understandable parts. I do not feel diminished by my kinship to Turing's machine.

....

CHAPTER 5

ALGORITHMS AND HEURISTICS

When I was an undergraduate at MIT, one of my roommates had several dozen pairs of socks, each pair with a slightly different color or design. He frequently postponed doing his laundry until he was completely out of clean socks, so whenever he washed them he had the not inconsiderable task of matching them all up again in pairs. Here is the way he would do it: First, he would pull a random sock out of the pile of clean laundry, then he would extract another sock at random and compare it to the first to see if it matched. If it didn't, he would throw the second sock back and pull out another one. He would keep doing this until he found a match, and then he would go through the same sequence all over again with a new sock. Since he had to look though a lot of laundry, the process went very slowly—especially at the beginning, because there were a lot more socks to be examined before a match turned up.

He was studying for a degree in mathematics, and was apparently taking some kind of course in computers. One day when he had hauled his laundry basket back to our rooms, he announced, "I have decided to use a better algorithm for matching my socks." What he meant was that he was now going to use a procedure of a fundamentally different nature. He pulled out the first sock and set it on the table, then he pulled out the next sock and compared it with the first sock; since it didn't match, he set it next to the first. Now each time he pulled out a sock he would compare it to the growing row of socks on the table. When he found a match, he would bundle that pair together and throw them in his sock drawer. When he didn't, he would add the unmatched sock to the row. Using this method, he was able to pair up his socks in a small fraction of the time it had previously required. His parents, who had paid a great deal of money for his college education, would have been proud to know that he was putting his newfound learning to such practical use.

THE ALGORITHMIC GUARANTEE

With or without socks, an *algorithm* is a fail-safe procedure, guaranteed to achieve a specific goal. The word "algorithm" comes from the name of the Arabian mathematician al-Khwarizmi, who wrote down an extensive collection of algorithms in the ninth century. The word "algebra," in fact, comes from *al jabr* ("the transposition"), a term in the title of one of his books. Many of al-Khwarizmi's algorithms are still used today. He described them, of course, in Arabic, which may be why this language gained a reputation as the language of magic spells. (It has even been suggested that the incantation "abracadabra" is a corruption of al-Khwarizmi's full name, Abu Abdullah abu Jafar Muhammad ibn Musa al-Khwarizmi.)

Computer algorithms are usually expressed as programs. Since the term refers to the sequence of operations rather than the particular way they are described, it is possible to express the same algorithm in many different computer languages, or even to build it into hardware by connecting the appropriate registers and logic gates.

Usually, many algorithms can compute the same result. As in the sock example, different algorithms require different amounts of time to complete a given task. Certain algorithms may also offer other kinds of advantages: they may use only a small amount of a computer's memory, or they may require a particularly simple pattern of communication that is easy to wire into its hardware. The difference in speed and memory requirements between a good algorithm and a bad one is often a factor of thousands or even millions. Sometimes the discovery of a new algorithm allows you to solve previously intractable problems.

Because an algorithm can be implemented in many different ways and can be applied to problems of varying size, you can't judge how fast an algorithm is by measuring the time that elapses before a solution to your particular problem is reached. The time will vary with the method of implementation and the size of the problem. Instead, we usually describe the speed of an algorithm by how much the time required to complete the task grows along with the size of the problem. In the sock-pairing example, most of the time is spent in pulling the socks out of the basket, so we can compare the two sock algorithms by asking how the number of socks pulled out in each one compares to the total number of socks. Assume that there are n socks in the laundry basket. In the first algorithm, finding two that match requires pulling out and putting back an average of half the remaining socks, so the number of sock removals is proportional to the square of the number of socks. In analyzing algorithms, we usually don't bother to calculate the exact constant of proportionality; instead, we simply say that the algorithm is *order n^2*, meaning that for large problems the time required for completion grows as the square of the problem size. This means that if there are ten times as many socks, the first algorithm will take a hundred times as long, so it is not a very good algorithm to use for pairing large numbers of socks. In the second algorithm, however, each of the n socks is pulled out only once, so the algorithm is order n. When you use the second algorithm to sort ten times as many socks, the task will take only ten times as long.

One of the greatest joys in computer programming is discovering a new, faster, more efficient algorithm for doing something—particularly if a lot of well-respected people have come up with worse solutions. Computer scientists can gain fame and admiration—at least among other computer scientists—by discovering a faster algorithm for a common problem. Since a bad algorithm can take weeks to solve a problem that a good algorithm can solve in minutes, the classical form of algorithmic one-upmanship is to write a new program and compute the right answer while your colleague's inferior program is still running.

Often the best algorithm is not obvious. Consider the problem of sorting a deck of sequentially numbered cards into ascending order. One way to do this is to start by looking through the entire deck for the lowest-numbered card. This card is removed and it becomes the first card of the sorted output pile. Next we look for the lowest card among those that remain. This in turn is removed and placed on top of the first card. The process is repeated until the unsorted cards are exhausted and the deck is arranged in ascending order. This procedure requires looking through the entire deck each time a card is extracted. Since there are n cards, each of which requires n comparisons, the run time of the algorithm is order n^2.

If we know that the cards are sequentially numbered from 1 to n, then we can sort them by a different method—one that uses a recursive definition, like the Logo procedure for drawing a tree described in chapter 3. To sort a deck of cards recursively, go through the deck once, moving cards with a value lower than the mean to the bottom half of the deck and leaving the cards with higher-than-average value in the top half. Then sort the two halves of the deck using the same algorithm. Applying the same algorithm recursively to each half of the deck will involve applying it recursively to each half of the half-decks, and so on. Each step of the recursion halves the number of cards to be sorted; the recursion ends when there is only one card—in which case, it is already sorted. Because this algorithm involves repeatedly dividing the ca 's until you are examining only one, it will require a time proportional to the number of times n cards can be divided—or the logarithm to the base 2 of the number of cards. So the order of this algorithm is $n \log n$. (If you don't remember what logarithms are, never mind. They are all small numbers, so they can be safely ignored.)

There is an even more elegant recursive algorithm, which doesn't require the cards to be sequentially numbered; it would be useful for putting a large number of business cards into alphabetical order, for example. This algorithm, called *merge sort*, is harder to understand, but it's so beautiful that I cannot resist describing it. The merge-sort algorithm depends on the fact that it's easy to merge two already sorted stacks into a single sorted stack by successively pulling top-ranked cards off the top of one or the other of the stacks; this merge procedure is a subroutine of the algorithm, and the algorithm works like this: If your stack consists of only one card, then that card is already sorted. Otherwise divide the stack in half, and recursively use the merge-sort algorithm by sorting each half and then combining them using the merge procedure described above. That's all there is to it. (If this sounds too simple to work, you might want to try it with a few cards. Start with eight.) The merge-sort algorithm is a good example of the almost mysterious power and elegance of recursion.

A sorting algorithm that requires just $n \log n$ steps, like merge sort, is pretty fast. In fact, it is about the fastest algorithm possible. Proving that particular statement is beyond the scope of this book, but the reasoning that underlies the proof is interesting. It can be shown by counting the number of possible orderings of n cards. From this count, it is possible to compute that $n \log n$ bits of information must be known in order to put the cards in the correct order. Since each comparison of two cards produces only 1 bit of information (either the first card is greater than the second, or it is not), then to sort n numbers requires at least $n \log n$ comparisons, so in this case the merge-sort algorithm is about as good as any other. Books have been written on the topic of choosing the proper sorting algorithm. In many cases, where certain constraints are put on the sorting, or particular knowledge is available about the objects being sorted, the fastest possible sorting algorithm remains unknown. Still, on the scale of tasks for which we would like to design algorithms, sorting is considered relatively easy.

An example of a difficult task is called the *traveling salesman problem*. Imagine that a traveling salesman has to visit n cities. Given the traveling distance between each of the cities, in what order should the salesman visit the cities to minimize total distance traveled? No one knows an algorithm that is order n^2, or order n^3, or n to any power, which will accomplish this. The best algorithm known is order 2^n, meaning that the time required grows exponentially with the size of the problem. If we add ten more cities to the salesman's itinerary, the problem gets a thousand times harder ($2^{10} = 1,024$). If we add thirty more cities, it gets about a billion times harder (2^{30} is about 10^9). Exponential algorithms are not much use when problems get large, but for the traveling salesman problem they are the best algorithms we know. The fastest computer in the world, working for billions of years, would not have enough time to find the best route for just a few thousand cities.

The traveling salesman problem may seem unimportant, but it turns out to be equivalent to a lot of other problems—the so-called *N-P complete problems* (N-P stands for "nondeterministic polynomial")—that it would be very useful to solve. A fast solution to the traveling salesman problem would lead immediately to a solution of these additional problems: for example, certain codes used for protecting secret information would become easy to break. Anyone who uses these codes is betting that no fast algorithm for solving the traveling salesman problem will ever be found. It's probably a safe bet.

No predictable technical breakthroughs in computers will help solve the traveling salesman problem, since even a computer a billion times faster will still be stumped by the addition of a few more cities. Exponential algorithms are just too slow to use for large problems. What could make a difference is the invention of a new algorithm: no one has ever proved that a fast algorithm for the traveling salesmen problem *cannot* exist. The study of algorithms has made significant progress in just the last few decades, and finding a fast one for the traveling salesman problem—or else proving that a fast one does not exist—remains one of computing's holy grails.

SETTLING FOR ALMOST ALWAYS

As hard as the traveling salesman problem is, it is not one of the most difficult problems to solve on a computer. Some problems are known to require even more than exponential time to solve. As discussed in the previous chapter, there are noncomputable problems that we know no algorithm can solve. Even when algorithms exist for certain problems, they are not necessarily the best approach. An algorithm, by definition, is guaranteed to get the job accomplished, but this guarantee of success often comes at too high a price. In many cases, it is more practical to use a procedure that only *almost always* gets the right answer. Often, "almost always" is good enough. A rule that tends to give the right answer, but is not guaranteed to, is called a *heuristic*. It is often more practical to use a heuristic than an algorithm: for instance, there are many effective heuristics for the traveling salesman problem—procedures that will provide an almost optimal route very quickly. In fact these heuristics usually do find the best route, although they are not absolutely guaranteed to do so. A real-life traveling salesman would presumably be happier with a good, fast heuristic than with a slow algorithm.

A simple example of the use of heuristics is the game of chess. A talented programmer who is only an average chess player can write a chess-playing program that will consistently beat the programmer. Such a program is not an algorithm, because it is not guaranteed to win. Heuristics make educated guesses; good heuristics usually make the right guess. Some of the most impressive behaviors of computers are the result of heuristics rather than of algorithms. (Philosophers have written a great deal of nonsense about "the limitations of computers" when what they are really talking about are the limitations of algorithms.)

A good chess-playing program can be written based on the following heuristics:

1. Estimate the relative strength of each player's position by counting the number of pieces of each type remaining on the board.
2. Move so as to put yourself in the strongest possible position a few moves in the future.
3. Expect your opponent to adopt a strategy similar to your own.

Each of these rules is only an approximation of the ideal strategy, and it is possible to imagine situations in which each is actually wrong. The relative strength of a player's position, for example, depends not just on the number of pieces but also on their position. A good position can often be more advantageous than an extra piece. Regardless, the first heuristic is generally correct; in most cases, having more pieces is better. Even before computers, chess players developed a simple method of numerically scoring the relative strengths of two players' positions by assigning one point for a pawn, three for a bishop, five for a rook, and so on, and using the total score of each player's remaining pieces as a measure of strength.

Based on these heuristics, you can write a chess-playing program that will trace out all plausible lines of play for the next few moves. Of course, it would be preferable if the program considered *all* lines of play, plausible or not, *all* the way to the end of the game. This was easy in the game of tic-tac-toe, but where chess is concerned it is impractical even for the fastest computers. In a typical midgame chess position, a player has about thirty-six potential legal moves, each of which leads to thirty-six possible responses by the opponent. Since the average chess game lasts for more than eighty moves, the computer would have to search something on the order of 36^{80} possibilities, or about 10^{124} possibilities. Such a search could not be accomplished by the fastest modern computers in hundreds of years. The problem is that the possible lines of play grow exponentially with the number of moves; it is thus impractical to look more than about five to ten moves ahead—which is why computers use the heuristics I have just listed for evaluating their moves.

Let us for the moment accept the second heuristic as correct and agree that the best line of play is the one that optimizes a player's position a few moves into the future. Let us further specify that the chess-playing program will look six moves ahead. According to the first heuristic, the program will evaluate the strength of both players after the sixth move by counting the number of pieces on each side remaining on the board and scoring them using the point system I have described. The relative strengths of the two players in any position considered will be judged by the difference between those scores.

Given these assumptions, what is the best way for the program to choose its next move? It's not enough for the computer to choose a move leading to the most favorable sequence of six future moves, because every alternate move in that sequence will be determined by the opponent. Instead, we must assume that the opponent will always try to choose a line of play that will favor the opponent's relative position; this is the assumption embodied in the third heuristic. To predict the opponent's line of play, the computer must place itself in the position of the opponent. The computer chooses a move by evaluating each legal move that can be made by its own side—white, say. The procedure for evaluating a possible move to be made by white depends on invoking a procedure for evaluating a possible move to be made by black, and vice versa. In effect, the computer follows every possible line of play for six moves, alternately putting itself in the position of black and white. The program tries out the moves on an imaginary board inside the memory of the computer, in much the same way that a chess master imagines lines of play "inside his head." The programs evaluating white's and black's positions call each other as subroutines, recursively. The recursion terminates after six levels, when the computer evaluates the score by counting the pieces.

Most chess-playing programs incorporate additional heuristics to abort searches of implausible lines of play and to search deeper in branches that involve the exchange of pieces. There are also more elaborate systems for evaluating positions without searching—for example, systems that award points for keeping control of the center or for protecting the king. Each of these heuristics is just an additional guess, and each one can improve the search in some situations at the cost of potentially making mistakes in others. With various refinements, this basic search procedure is at the heart of nearly every chess-playing program. It is effective because it takes advantage of the speed of the computer to consider many millions of alternate lines of play. Among these many millions of alternatives, there is often a variation that will surprise the programmer, or even an experienced human chess player. This ability to surprise is what allows the machine to play a better game of chess than the programmer.

Chess-playing machines have a long and sometimes dishonorable role in the history of computing. The eighteenth-century Hungarian inventor Wolfgang von Kempelen captured the world's imagination with a chess-playing automaton in the form of a mechanical turbaned Turk. As it turned out, the machine worked only because a chess-playing midget was hidden inside it. In 1914, the Spanish engineer Luis Torres y Quevedo built a mechanical device that played a simplified game of chess without the help of a concealed human being, and in the late 1940s Claude Shannon described how a computer could be programmed with a set of chess-playing heuristics similar to those listed here. Still, it was many years before computers were fast enough to play a decent game of chess, which comforted not a few philosophers who argued that chess playing was an example of the unique powers of the human mind. Modern computers, using the same heuristics, can now beat the best human chess players in the world (witness the victory in 1997 of IBM's Deep Blue over Garry Kasparov), so the philosophers have shifted the argument to other domains.

The simple search heuristics work because there are a relatively small number of responses to consider for each move. In checkers, where there are even fewer possible responses to each move, machines based on heuristics began beating the best human players in the 1960s. In the Chinese/Japanese game of Go, on the other hand, humans still reign, because the larger board (19 x 19) affords far more possible moves. (I prefer playing Go to chess, precisely because a search is less useful; it makes my impatience less of a disadvantage.)

FITNESS LANDSCAPES
••••••••••••••••••••••••

The use of heuristics to search through a set of possibilities is ubiquitous in computer programming and has applications far more important than game playing. This is often the way computers find "creative" solutions to problems—usually problems whose solutions are known to be among a large but finite set of possibilities called a *search space*. The search space in chess is the set of all possible lines of play; the search space in the traveling salesman problem consists of all possible routes linking the cities on the salesman's list. Since these spaces are too large to search exhaustively, heuristics are used to reduce the area to be searched. In the case of small search spaces, such as in the game of tic-tac-toe, the exhaustive search is preferable, because it is guaranteed to find the right answer.

Generally, the reason that a search space is large is because the possibilities are produced by forming combinations of simpler elements—the individual moves in chess, the city-to-city hops in the traveling salesman problem. This combining of elements leads to a *combinatorial explosion* of possibilities—an explosion that grows exponentially with the number of elements being combined. Since the possibilities are built from combinations of elements, there is a sense of distance in the space; combinations that share common elements are "closer" than the combinations that do not. This is why it is called a "space" and not just a set of possibilities. To extend the analogy, we can imagine the possibilities as lying in a two-dimensional landscape sometimes known as a *fitness landscape*. The desirability, or score, of each possible solution is represented by the altitude of a point in the landscape. If similar possibilities have similar scores, then nearby points will have similar altitudes, so the landscape will have well-defined hills and valleys. In this analogy, finding the best solution is like finding the top of the highest hill. Taking the traveling salesman problem as an example, we can imagine each point in the landscape as representing a particular travel itinerary for the salesman. The height of each point represents the distance the salesman must travel, with points representing efficient travel itineraries at higher altitudes. The best itinerary will be at the top of the highest hill.

One of the simplest ways of searching such a space is to compare points at random and remember the best one found.

The number of points that can be searched this way is generally limited only by the amount of time available, and the procedure can be applied to any type of space. It is the equivalent of parachuting scouts into various locations in the landscape and asking them to report back their altitude. It is not a very efficient way to find the top of a hill. If the space is large, then in any practical amount of time only a tiny portion of the possibilities will be investigated, and therefore the best point found is unlikely to be one of the highest.

In a search space like that of the traveling salesman problem, where nearby points are likely to have similar scores, it is usually better to use a procedure that searches a path through the space by traveling from point to nearby point. Just as the best method for finding a peak in a hilly landscape is to walk uphill, the equivalent heuristic is to choose the best of nearby solutions found in the search space. In the traveling salesman problem, for example, the computer might vary the best-known solution by exchanging the order of two of the cities in the itinerary. If this variation leads to a more efficient tour, then it is accepted as a superior solution (a step uphill); otherwise, it is rejected and a new variation is tried. This method of search will wander through the space, always traveling in an uphill direction, until it reaches the top of a hill. At this point, the solution cannot be improved by exchanging any pair of cities.

The weakness of this method, which is called *hill climbing*, is that although you thereby reach the top of a hill, it is not necessarily the highest hill in the landscape. Hill climbing is a heuristic, not an algorithm. There are other heuristics similar to hill climbing which are less likely to get you stuck on top of one of the foothills. For instance, you could repeat the hill-climbing process many times, starting from different random locations (that is, you might ask the parachutists to climb uphill). Or you could occasionally take a step downward to avoid getting stuck. There are many such variations, each with its own advantages and disadvantages.

Heuristics like hill climbing work well on the traveling salesman problem and produce good answers in a short amount of time. Even when thousands of cities are involved, it is usually possible to find at a good solution to the problem by starting out with a reasonable guess and improving it by hill climbing. So why is the traveling salesman problem considered so difficult? Using heuristics, we can *almost* always get *almost* the best itinerary. But a method that *almost* always works is not an algorithm. Periodically, a great deal of fuss is made by someone who has "solved" the traveling salesman problem; so far, all anyone has actually done is come up with a new heuristic. Fast heuristic solutions to the traveling salesman problem are not difficult to dream up; it is finding a fast algorithm that is the difficulty.

There are many problems for which we do not need exactly the right answer every time—problems for which we can accept a less-than-perfect solution. Even when we want a perfect answer, we may not be able to afford it. For such problems, computers can produce an educated and well-considered guess. Because the computer is able to consider an enormous number of combinations and possibilities, such a guess will often surprise the programmer. When a computer uses heuristics, it is capable both of surprises and mistakes—which makes it a little more like a person and a little less like a machine.

CHAPTER 6

···

MEMORY: INFORMATION
AND SECRET CODES

So far, we have mostly ignored the limitations imposed on a computer by the size of its memory. An idealized universal computer has an infinite memory, but a real computer's memory is limited, usually by expense. As long as the size of the memory is adequate for the task at hand, we are free to ignore the limitation, but some memory-intensive algorithms and applications store such large amounts of data that the amount of memory available becomes an important consideration. Applications that manipulate representations of the physical world—such as images, sounds, or three-dimensional models—are often memory-intensive. Knowing how much memory is required for a given application is important not only in judgments about whether or not the computer is big enough to handle it but also for estimating the time required to process the information.

The bit—the unit of measure for information—is appropriate both for the communication of information and for its storage. In a sense, communication and storage are just two aspects of the same thing: communication sends a message from one place to another; storage "sends" a message from one time to another. Unless you are accustomed to thinking in four-dimensional spacetime terms, this equivalence between moving and storing may seem strange, but think of mailing a letter as a means of communication which has aspects of both. Mailing a letter to someone else is a way to move information in space; mailing a letter to yourself is a way to store information in time. When examined closely, any form of communication is seen to have both a spatial and a temporal aspect. One way that electronic computers store information is to constantly recirculate it, in the electronic equivalent of a self-addressed letter.

We know that computers with n bits of memory can hold up to n bits of information, but how do we determine how many bits are required to represent a given piece of information? For instance, how many bits are in the words of this book? Calculating the answer turns out to be not particularly easy; in fact, there are several different correct answers. Thinking about this question leads us to ideas about compression, error detection and correction, random numbers, and secret codes.

The number of bits required to send or store a given piece of data will depend on how the data is encoded. One way to represent a complex message, like the text of a book, is to represent it as a sequence of simpler parts: for example, the characters that spell out the text. In this common representation, the number of bits in the message is equal to the number of characters in the text times the number of bits per character. The text of this book contains about 250,000 characters, and my computer uses a code that requires 8 bits (a byte) to store a character, so the size of the file that the computer uses to store the text is about 2 million bits. You may be tempted to conclude that this book contains about 2 million bits of information, since that's how much memory the computer uses to store the text, but this is only one measure of the information—a measure that depends on the representation of the message. It's a useful measure, because it tells you not only how much memory the computer needs to store the information but also how much time is needed to process it. For example, if I know that my computer can write information on a disk at 20 million bits per second, and I know that it uses a 2-million-bit representation of this book, then I can calculate that it will take about 1/10 of a second to store the book in a file on disk.

The problem with using the number of characters times 8 as a measure of the number of bits in the text is that it depends on the representation scheme used by the computer. A different computer, or a different application program running on the same computer, might store the very same sequence of characters in a different number of bits. For instance, with 8 bits per character, it is possible to represent 256 different characters, but the text in this book uses fewer than 64 different characters—26 letters, upper and lower case, numerals, and punctuation marks. Therefore, a more efficient code would represent each character by using 6 bits of information ($2^6 = 64$), and hence compress the representation of the text to only 1.5 million bits.

It would be nice to have a measure of information that didn't depend on the form of representation. A more fundamental measure of information would be the *minimum* number of bits required to represent the text. This is easy to define, but not necessarily easy to calculate.

COMPRESSION

How much can we compress a given text without losing information? Reducing the number of bits per character from 8 to 6 is a simple form of compression. Other compression methods are based on taking advantage of regularities in the text. For instance, the letters T and E occur far more often in most English-language texts than the letters Q and Z. A more efficient code would use a shorter sequence of 1's and 0's to represent the more common letters. Using a variable-length encoding of characters to achieve a more compact representation was a trick used by early telegraph operators and radio amateurs. In Morse code, the letter E is represented by a single dot and the letter T by a single dash. Less common letters, like Q and Z, are represented by sequences of up to four dots and dashes. Because a third type of signal—a space—is used to mark the ends of letters, the dots and dashes of Morse code don't exactly correspond to 1's and 0's, but the principle is similar.

To make a variable-length code using 1's and 0's, it is necessary to choose the code patterns carefully, so that a stream of bits can be broken unambiguously into characters. This is possible as long as no bit sequence used to represent a character begins with a subsequence of 1's and 0's used to represent another character. For example, all common characters could be represented by 4 bits, starting with a 1, while less common characters might have 7 bits and start with a 0. This would allow the stream of bits to be divided unambiguously into short and long characters. Choosing a variable-length code that best takes advantage of the relative frequency of the various letters will result in substantial compression of the text. In the case of the text of this book, it would reduce the number of bits from the original 2 million to about 1 million, a compression of 50 percent.

Any method of compressing takes advantage of regularities in the data. The code just described takes advantage of regularities in the rate of occurrence of single characters, but there are other regularities that can be exploited. For instance, not all pairs of adjacent letters occur with equal frequency in this book. The letter Q is almost always followed by the letter U, and the letter Z is never followed by the letter K. By developing a system of variable-length code for pairs of letters rather than for individual letters, we can take advantage of the fact that two-letter combinations do not occur with equal frequency. The code can use shorter sequences of bits for the more common pairs and long sequences for pairs that occur rarely. If we use this method, the number of bits required to store the text of this book could be reduced by a further 10 percent, to an average of about 3.5 bits per letter.

A still more efficient code would take advantage of regularities that occur in longer sequences of letters. For example, the word "the" occurs about 3,000 times in this text. It would be advantageous to use a code that uses a relatively short sequence of bits to represent this entire word. Similarly, there are many other words, such as "computer" and "bit," that occur so often in this particular text that they, too, are worth encoding.

There are also regularities beyond these statistical regularities in the letter sequences. For instance, there are regularities in grammar, sentence structure, and punctuation that allow further compression of the text. But at some point we will begin to get diminishing returns. In the end, the compression that uses the best available statistical methods would probably reach an average representation size of fewer than 2 bits per character—about 25 percent of the standard 8-bit character representation.

Compression works fairly well on text, but it works even better on signals that are representations of the real world, like sounds and pictures. These signals are usually read into the computer by a process known as *analog-to-digital conversion*. Such inputs—the intensity level of a sound, say, or the brightness of a light—are usually continuously variable, analog signals. For example, a dot, or *pixel*, in a black-and-white photograph may be either white or black or any of the infinite shades of gray in between. Since the computer has no way of representing an infinite number of possibilities, it simplifies the signal by reducing each pixel to one of a finite set of levels of gray. Typically, the number of gradations is an exact power of 2, so that it will fit into a specific number of storage bits. For instance, the brightness of a dot in a black-and-white picture might be represented by 8 bits, so that 256 shades of gray could be represented. A higher-quality image would be represented with a 12-bit code, producing 4,096 shades of gray. A color image might use 24 bits per dot—8 bits for the intensity of each of the three primary colors.

The other parameter determining the quality of a photographic image is its *resolution*—that is, the number of pixels used to represent it. A high-resolution image produced by an array of 1,000 x 1,000 dots will be a more accurate representation than an image with a resolution of 100 x 100. However, since the representation of the high-resolution image uses 1,000,000 pixels instead of 10,000, your computer will need 100 times as much memory for storage, and processing the image will take 100 times as long. Quality costs.

Since high-resolution images can contain a large number of bits, it is often desirable to compress them, in order to reduce the cost of storage and transmission. This is especially true of moving images, which typically contain from 24 to 100 frames per second. Fortunately, images are relatively easy to compress, because there is a high degree of regularity in an image. In most pictures, the intensity and color of a particular pixel is often nearly identical to the intensity and color of neighboring pixels. Two pixels representing adjacent parts of the same cheek in the image of a face, say, would likely be very similar in brightness and color. Most image-compression algorithms take advantage of this similarity. An image-compression algorithm may represent areas of uniform brightness and color by having just a few numbers represent the color and size of the area. Other image-compression methods take advantage of more complex forms of regularity: for example, similar textures in different parts of the image. For moving pictures, such as television broadcasts, compression methods generally take advantage of the similarity of sequential frames. Using such techniques, one can often compress the representation of a photograph by a factor of 10 and of a moving image by a factor of 100. Similar compression methods can be applied to sounds.

These compression methods lead to a counterintuitive notion of the amount of information contained in a picture. If the minimum number of bits required to represent the image is taken as a measure of the amount of information in the image, then an image that is easy to compress will have less information. A picture of a face, for example, will have less information than a picture of a pile of pebbles on the beach, because the adjacent pixels in the facial image are more likely to be similar. The pebbles require more information to be communicated and stored, even though a human observer might find the picture of the face much more informative. By this measure, the picture containing the most information would be a picture of completely random pixels, like the static on a damaged television set. If the dots in the image have no correlation to their neighbors, there is no regularity to compress. Such pictures look completely meaningless to us—and may truly be meaningless—but they require the greatest amount of information in order to be represented by a computer.

The minimal representation measure of information does not correspond well to our intuition about information content because the computer is making no distinction between meaningful and meaningless information. The computer must represent the color of every pixel, or the position of every pebble on the beach, even though these details may not be important to the viewer. Deciding what information is and is not meaningful is a subtle art; it depends on how an image is being used and who is using it. The position of a tiny blemish on an X-ray image might be irrelevant to an untrained eye but very meaningful to a physician. A great artist like Picasso may be able to "compress" the image of a complex scene into a few simple lines, but in doing so he exercises a complex judgment in deciding which aspects of the image communicate the most meaning (see Figure 23).

If the computer were to compress an image by storing only the meaningful information, then the number of bits in the representation would correspond more closely to our

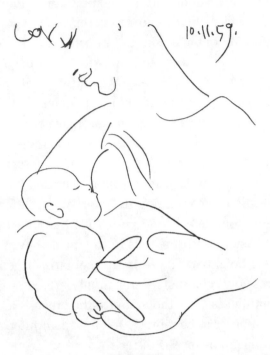

FIGURE 23

Picasso sketch

commonsense notion about the information contained in the image. For example, the computer might represent the array of random pixels by indicating that this image has no regularity and no meaningful information. When asked to reconstruct this picture, it could simply generate another array of random pixels. The details—which pixel was which exact shade—would be different in the original and the reconstructed images, but these differences would be meaningless to the human eye.

Many image- and sound-compression algorithms discard certain meaningless information in order to reduce the size of the representation. These so-called *lossy* compression algorithms assume that a certain level of detail in the image or sound will be ignored by the eye or the ear. Lossy compression methods are generally used when it is known that the decompressed information will serve some particular purpose. For instance, if a particular detail appears only in a single frame of a movie, it may be safe to throw it away, since it will go unnoticed.

There is another important form of image representation, which is able to achieve an even greater degree of compression than the methods just described. If the process that generates the original image is known, then it may be more efficient to store a record of that process rather than a record of the image itself. For instance, if the image was a drawing, created by a person who drew a series of lines, then the drawing may be represented by storing a list of the lines—a representation scheme often used by computers for making simple line drawings.

The idea of representing something by storing the procedure or program that generated it is applicable to other types of data as well, such as sounds. Where sound is concerned, the technique may seem no more profound than the notion of recording a piece of music by writing down the score, but in a computer the score is able to represent every detail necessary to reproduce the original—the tuning of the instruments, the bowing of the violins, even the mood of the orchestra. If an object can be generated by a computer, then by definition there is a precise procedure for generating it, so a description of that procedure will serve as representation of the object.

This leads us to another measure of information: *The amount of information in a pattern of bits is equal to the length of the smallest computer program capable of generating those bits.* This definition of information holds whether the pattern of bits ultimately represents a picture, a sound, a text, a number, or anything else. The definition is interesting, because it allows for any type of regularity within the pattern. In particular, it subsumes all the methods of compression described above. (It might seem that such a definition depends on the details of the computer's machine language, but recall that any computer can simulate any other, so the measure from one computer to another will vary only by the small amount of code needed to perform the simulation.)

Once a string of information is compressed as much as possible, it will exhibit no regularity. This is true because any regularity would be an opportunity for further compression. The string of 1's and 0's representing optimally compressed text would look completely random, like the record of the flipping of a coin. In fact, many mathematicians use this property of incompressibility as a definition of randomness—a satisfyingly simple definition, but one that often is not very useful in practice, since it's very difficult to tell whether or not a given string of bits is random in this sense. It's easy to decide, when we recognize any regularity, that a string can be compressed—but we can't prove that the string *cannot* be compressed if we see no pattern. The pseudorandom number sequences described in chapter 4 are a good example of sequences that appear random but have an underlying pattern. By the above definition of randomness they are highly nonrandom, because a very long sequence can be briefly described, simply by describing the algorithm that produced it—in this case, the roulette-wheel simulation.

ENCRYPTION

Those sequences that appear random but have a hidden underlying pattern can be used to create codes for *encrypting* data. Imagine, for example, that I wish to send a secret message to a friend of mine. If we both have a copy of the same random-number generator, so that we can generate the same sequence of pseudorandom numbers, we could use this sequence to hide the content of the message from anyone who might intercept it. Let's say that the message to be transmitted is a stream of bits representing characters, using the standard 8-bit-per-character representation. This standardized representation could presumably be interpreted and understood by any eavesdropper; it is what cryptographers call the *plain text* of the message. To encrypt the message, we pair each bit in the plain text to the corresponding bit in the pseudorandom bit stream. If the pseudorandom bit is 1, we invert the corresponding plain-text bit. If the pseudorandom bit is 0, we leave the corresponding plain-text bit alone. This will invert about half the bits in the plain text, but the eavesdropper won't know which half. Unless the eavesdropper knows the pseudorandom sequence, this string of 1's and 0's will be utterly meaningless. My friend the recipient, on the other hand, knows how to generate exactly the same random sequence, which can be used to reinvert the inverted bits, thereby restoring (decrypting) the message. This method, or something very similar to it, is at the heart of most encryption schemes.

Encrypting a message is analogous to sending it in a locked box that can be opened only with a special key. In the encryption method just described, the key is the random-number generator. Anyone who has the key is able to perform the conversion. In the example above, the same key is used for encryption and decryption, but it is also possible to construct codes that use different keys for encryption and decryption. In a public encryption scheme, the keys for encryption and decryption are different, and an eavesdropper who knows the encryption key will not thereby know the key needed to decrypt. This method of transmitting messages is extremely useful. For example, if I wish to receive an encrypted message, I can publicize the description of the key necessary to encode messages to me. Anyone will be able to send me a secret message, whether I know them or not. Since the public key tells the sender only how to encrypt the message, not how to decrypt it, others will not be able to unlock the encoded message. Only my private key, which I keep secret, allows an encrypted message to be converted back into plain text. This is called *public key encryption.* Public key encryption solves an important practical problem: for example, many businesses that accept credit card numbers over the Internet publish their own public key, so that customers can encrypt their credit card numbers without fear that they will be intercepted and read.

The public-key-encryption scheme is also useful in reverse, for authenticating messages. In this case, I publicize the key for decrypting a message but keep secret my key for encryption. Whenever I want to send out a message that I wish to "sign" as being verifiably from me, I encrypt the message with my private key. Any recipient of the message can use the public key for decrypting the message. They will know it was really from me because only someone who knows my private key could have encrypted such a message.

ERROR DETECTION
....

Encoding and decoding bits have many other applications besides compression and security. For instance, there are situations in which we may represent a message using more bits than necessary, in order to reduce the chance of error. Codes that use some form of redundancy for detecting transmission errors—for example, a 0 that was received as a 1— are called *error-detecting codes*. Other codes, called *error-correcting codes,* contain enough redundant information to correct as well as detect such errors.

An obvious form of redundancy is to send a message more than once. Sending a message twice allows for error detection. If two copies of the same message are transmitted, but slightly different messages are received, then there must have been some error in transmission. A simple error-*correcting* code would be to repeat the message three times. Assuming that only one of the messages was corrupted, then the recipient would reconstruct the correct message by choosing the two copies that were the same.

Fortunately there are error-detecting and error-correcting codes that achieve results with far less redundancy. A commonly used scheme for detecting errors is a *parity code*. This scheme can detect a single-bit error in a message of any length by the addition of one redundant bit. As a specific example of a parity code, consider the 8-bit code often used for transmitting characters over noisy communications lines. The eighth bit, called the *parity bit*, is 1 if, and only if, the number of 1's in the seven other bits is even. This means that the number of 1's in the 8-bit sequence should always be odd. If noise in the transmission line causes a 1 to be received as a 0, or vice versa, then the 8-bit message that is received will have an even number of 1's. The recipient therefore will be able to detect that there has been an error. Similar parity schemes are used for detecting errors in the memory systems of computers. One bit of parity can be used to detect an error in a message of any number of bits. A limitation of this simple parity code is that it's good at detecting only a single error. A message that has two inverted bits will have the correct parity, even though the data is incorrect.

By using multiple parity bits, it is possible to detect multiple errors. It is also possible to give the recipient enough information not only to detect an error but to correct it; that is, the recipient is able to reconstruct the original message even though there has been an error. An example of such a code is the two-dimensional parity code illustrated in Figure 24.

This code contains 9 bits of message information and 6 bits of parity. The message bits are arranged in 3 rows of 3 bits each. There is 1 parity bit for every horizontal row and 1 for every vertical column. A single-bit error in a message will cause two parity failures to be detected, one in a row and one in a column. The recipient of the message will then know that the bit at the intersection of the failing row and the failing column is incorrect and should be inverted. If there is an error in transmission of one of the parity bits, on the other hand, then either a row or a column will show an incorrect

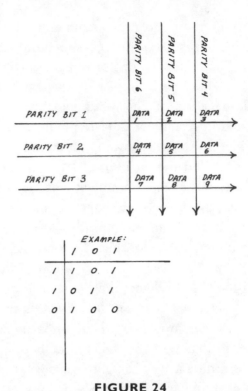

FIGURE 24

An error correction code with 9 data bits and 6 check bits

parity but not both. The bits are drawn in a two-dimensional pattern to help you visualize the structure of the code, but they can be transmitted in any order. Such error-correcting codes are often used to protect each word in the memory of a large computer. Using similar techniques, many other codes can be constructed which will detect and/or correct various types and numbers of errors.

Error-correcting codes are able to deal with signaling errors that occur in transmission and storage of information, but what about errors in computation itself? It turns out that it is also possible to build logic blocks that produce correct answers even if some of the blocks from which they're constructed are operating incorrectly. Again, the basic tool is some form of redundancy. One way to build a fault-tolerant logic block is to copy each logic block three times. A majority-voting block, such as is shown in Figure 12 of chapter 2, can be used to combine the answers from the three copies. If one of the copies makes an error, it will be outvoted by the others. This simple method protects against any single error (except within the majority-voting block itself).

Given a well-defined set of possible errors—such as wires breaking, switches getting stuck, and 0's switching to 1's—it is possible to construct an arbitrarily reliable computing device out of arbitrarily unreliable components. This task simply requires using enough redundant logic in a systematic form. For instance, if you could build a new type of switching device—say, a molecular switch—that was extremely fast, or extremely inexpensive, but failed 20 percent of the time, you could still use it to build a computer that produced answers with 99.99999-percent reliability, by building the proper redundancy into the circuitry.

Does this mean that you can construct an arbitrarily reliable computer? Not exactly. While a computer can be constructed so as to eliminate a particular type of error, errors of unanticipated types may occur which produce correlated errors in the redundant module. For instance, the burnout of one module may cause another module to overheat, or some kind of magnetic pulse may even cause all modules to err simultaneously. Engineers are capable of designing a logic block that can deal with any type of error they can imagine, but the history of technology shows that our imaginations are not always sufficient. The most dramatic failures are usually surprises.

A second reason not to expect a perfect computer is that most computer failures are not caused by incorrect operation of the logic. They stem from errors in design—usually in the design of the software. Programmed computers, including their software, are by far the most complex systems ever designed by human beings. The number of interacting components in a computer is orders of magnitude larger than the number of components in the most complex airplane. Modern engineering methods are not really up to designing objects of such complexity. A modern computer can have literally millions of logical operations going on simultaneously, and it is impossible to anticipate the consequences of every possible combination of events. The methods of functional abstraction described in the previous chapters help keep the interactions under control, but these abstractions depend on everything interacting as expected. When unanticipated interactions occur (and they do), the assumption on which the abstraction rests breaks down, and the consequences can be catastrophic. In a practical sense, the behavior of a large computer system, even if no failures occur, is sometimes unpredictable—and this is the overarching reason that it is impossible to design a perfectly reliable computer.

....

CHAPTER 7

SPEED: PARALLEL COMPUTERS

Besides differing in the amount of memory, one type of universal computer can differ from another in the speed with which its operations are carried out. The speed of a computer is generally governed by the amount of time it takes to move data in and out of memory.

The sort of computers we have been talking about so far are sequential computers—they operate on one word of data at a time. The reason that conventional computers operate this way is largely historical. In the late 1940s and early 1950s, when computers were being developed, the switching elements (relays and vacuum tubes) were expensive but relatively fast. The memory elements (mercury delay lines, magnetic drums) were relatively cheap and slow. They were especially suitable for producing sequential streams of data. Computers were designed to keep the expensive processor switches as usefully busy as possible, while not placing too much demand on the speed of the memory. These early computers were room-size, with the expensive processor on one side and the slow memory on the other. A trickle of data flowed between.

As computer technology improved, software grew increasingly complex and expensive and it became increasingly difficult to train programmers, so in order to preserve the investment in software and training, the basic structure of computers remained unchanged. There was little motivation to rethink the two-part design, because technology was progressing at such a rapid rate that it was easy to build faster and cheaper machines just by re-creating the same kind of computer in a new technology.

The speed of computers tended to double every two years. Vacuum tubes were replaced by transistors and eventually by integrated circuits. Delay-line memories were replaced by magnetic-core memories, and then also by integrated circuits. The room-size machines shrank to a silicon chip the size of a thumbnail. Through all this changing technology, the simple processor-connected-to-memory design stayed the same. If you look at a modern single-chip computer under a microscope, you can still see the vestiges of the room full of vacuum tubes: one area of the chip is devoted to processing, another to memory. This is true in spite of the fact that the processor and memory parts of the computer are now made by the same methods, often on the same piece of silicon. The patterns of activity are still optimized for the earlier, two-part design. The section of the silicon chip that implements processing is kept very busy; the part that implements memory still trickles out data one word at a time.

The flow of data between processor and memory is the bottleneck of a sequential computer. The root problem is that the memory is designed to access a single location on each instruction cycle. As long as we stick with this fundamental design, the only way to increase the speed of the computer is to reduce the cycle time. For many years, the cycle time of computers could be reduced by increasing the speed of the switches: faster switches led to faster computers. Now this is no longer such an effective strategy. The speed of today's large computers is limited primarily by the time required to propagate information along their wires, and that in turn is limited by the finite speed of light. Light travels at about a foot per nanosecond (a billionth of a second). The cycle time of the fastest computers today is about a nanosecond, and it is no coincidence that their processors are less than a foot across. We are approaching the limits of how much we can speed up the computer without changing the basic design.

PARALLELISM
••••••••••••••••••

To work any faster, today's computers need to do more than one operation at once. We can accomplish this by breaking up the computer memory into lots of little memories and giving each its own processor. Such a machine is called a *parallel computer*. Parallel computers are practical because of the low cost and small size of microprocessors. We can build a parallel computer by hooking together dozens, hundreds, or even thousands of these smaller processors. The fastest computers in the world are *massively parallel* computers, which use thousands or even tens of thousands of processors.

As I have earlier described, computers are constructed in a hierarchy of building blocks, with each level of the hierarchy serving as the step to the level above. Parallel computers are the obvious next level in this scheme, with the computers themselves as building blocks. Such a construction can be called either a parallel computer or a *computer network*. The distinction between the two is somewhat arbitrary and has more to do with how the system is used than how it works. Usually a parallel computer is in one location, whereas a network is spread out geographically, but there are exceptions in both these rules. Generally, if a group of connected computers is used in a coordinated fashion, we call it a parallel computer. If the computers are used somewhat independently, we call the connected computers a computer network.

Putting a large number of computers together to achieve greater speed seems like an obvious step, but for many years the consensus among computer scientists was that it would work for only a few applications. I spent a large part of my early career arguing with people who believed it was impractical, or even impossible, to build and program general-purpose massively parallel computers. This widespread skepticism was based on two misapprehensions—one about how complex such a system would have to be, and the other about how the components of such a system would work together.

Scientists tended to overestimate the complexity of parallel computers, because they underestimated—or at least underappreciated—the rate of improvement in microelectronics-fabrication technology. It was not so much that they were ignorant of this trend as that the rate of technological change was unprecedented in history; it was thus extremely difficult for expectations and intuitions to keep abreast of the change. I remember giving a talk at a computer conference at the New York Hilton in the mid 1970s, in which I pointed out that current trends indicated that there would soon be more microprocessors than people in the United States. This was considered an outrageous extrapolation at the time. Although microprocessors had already been produced, the popular view of a computer was still that of a large set of refrigerator-size cabinets with blinking lights. In the question period at the end of the talk, one of my ill-disposed listeners asked, in a voice laden with sarcasm, "Just what do you think people are going to do with all these computers? It's not as if you needed a computer in every doorknob!" The audience burst out laughing, and I was at a loss for an answer, but as a matter of fact in that same hotel today, every doorknob contains a microprocessor that controls the lock.

Another reason that people were skeptical about parallel computers was more subtle, and also more valid. It was the perceived inefficiency of breaking a computation into many concurrent parts. This problem continues to limit the application of parallel computers today, but it is not nearly as serious a limitation as it was once thought to be. Part of the reason for the overestimation of the difficulty was a series of misleading experiences with early parallel machines. The first parallel digital computers were built in the 1960s by connecting two or three large sequential computers. In most cases, the multiple processing units shared a single memory, so that each of the processors could access the same data. These early parallel computers were usually programmed to give each processor a different task: for example, in a database application, one processor would retrieve the records, another processor would tabulate statistics, and the third would print out the results. The processors were used much as different operators on an assembly line, each doing one stage of the computation.

There were several inefficiencies inherent in this scheme, all of which seemed to grow along with the number of processors. One inefficiency was that the task had to be divided up into more or less independent stages. While it was often possible to divide tasks into two or three stages, it was difficult to see how to divide one task into ten or a hundred stages. As one detractor of parallel processing explained it to a newspaper reporter, "Well, two reporters might be able to write a newspaper article faster by having one of them gather the news while another writes the story, but a hundred reporters working on the article together would probably not be able to get it written at all." Such arguments were pretty convincing.

Another inefficiency stemmed from the shared access to a single memory. A typical memory could retrieve only one word at a time from a given area of memory, and this limitation on the rate of access created an obvious bottleneck in the system, which limited its performance. If more processors were added to a system already limited by its fetch rate, the processors would all end up spending more time waiting for data, and the efficiency of the system would decrease.

Moreover, processors had to be careful not to create an inconsistency by altering data that another processor was looking at. For example, consider an airline-reservation system. If one processor is working on the problem of reserving seats, it looks to see if a seat is empty and then reserves the seat if it is. If two processors are booking seats for two different passengers simultaneously, there is a potential problem: they may both notice that the same seat is empty and decide to reserve it before either has a chance to mark it as taken. To avoid this kind of mishap, a processor had to go through an elaborate sequence that locked out other processors from accessing data while that processor was looking at a data word. This further aggravated the inefficiencies attendant on the competition for the system's memory, and in the worst case it would reduce the speed of a multiprocessor system to the speed of a single processor—and even to less than the speed of a single processor. As noted, these inefficiencies worsened as the number of processors increased.

The final source of inefficiency appeared to be even more fundamental: the difficulty of balancing the tasks assigned to the various processors. To return to the assembly-line analogy: we can see that the rate of computation will be governed by the speed of the slowest step. If there is just one slow operation, the rate of computation is set by that single operation. It is not unreasonable to expect that in this case, too, the efficiency of the system will decrease as we increase the number of processors.

The best formulation of these inefficiency problems is known as Amdahl's Law, after Gene Amdahl, the computer designer who came up with it in the 1960s. Amdahl's argument went something like this: There will always be a part of the computation which is inherently sequential—work that can be done by only one processor at a time. Even if only 10 percent of the computation is inherently sequential, no matter how much you speed up the remaining 90 percent, the computation as a whole will never speed up by more than a factor of 10. The processors working on the 90 percent that can be done in parallel will end up waiting for the single processor to finish the sequential 10 percent of the task. This argument suggests that a parallel computer with 1,000 processors will be extremely inefficient, since it will be only about ten times faster than a single processor. When I was trying to get funding to build my first parallel computer—a massively parallel computer with 64,000 processors—the first question I usually got at the end of one of my pitches was "Haven't you ever heard of Amdahl's Law?"

Of course, I *had* heard of Amdahl's Law, and to tell you the truth I saw nothing wrong with the reasoning behind it. Yet I knew for certain, even though I couldn't prove it, that Amdahl's Law did not apply to the problems I was trying to solve. The reason I was so confident is that the problems I was working on had already been solved on a massively parallel processor—the human brain. I was a student at the Artificial Intelligence Laboratory at MIT, and I wanted to build a machine that could think.

When I first visited the MIT Artificial Intelligence Lab as a freshman undergraduate in 1974, the field of AI (as it was coming to be known) was in a state of explosive growth. The first programs that could follow simple instructions written in plain English were being developed, and a computer that understood human speech seemed just around the corner. Computers were excelling at games like chess, which had been considered too complicated for them just a few years earlier. Artificial vision systems were recognizing simple objects, like line drawings and piles of blocks. Computers were even passing calculus exams and solving simple analogy problems taken from IQ tests. Could general-purpose machine intelligence be all that far off?

But by the time I joined the laboratory as a graduate student a few years later, the problems were looking more difficult. The simple demonstrations had turned out to be just that. Although lots of new principles and powerful methods had been invented, applying them to larger, more complicated problems didn't seem to work. At least part of the problem lay with the speed limitations of computers. AI researchers found it unfruitful to extend their experiments to cases involving more data, because the computers were already slow, and adding more data just made them slower. It was frustrating, for example, to try to get a machine to recognize a pile of objects when recognizing a single object required hours of computing time.

The computers were slow because they were sequential; they could do only one thing at a time. A computer must look at a picture pixel by pixel; by contrast, a brain perceives an entire picture instantly and can simultaneously match what it sees to every image it knows. For this reason, a human being is much faster than a computer at recognizing objects, even though the neurons in the human visual system are much slower than the transistors in the computer. This difference in design inspired me, as it did many others, to look for ways of designing massively parallel computers— computers that could perform millions of operations concurrently and exploit parallelism more like the brain does. Because I knew that the brain was able to get fast performance from slow components, I also knew that Amdahl's Law did not always apply.

I now understand the flaw in Amdahl's argument. It lies in the assumption that a fixed portion of the computation, even just 10 percent, *must* be sequential. This estimate sounds plausible, but it turns out not to be true of most large computations. The false intuition came from a misunderstanding about how parallel processors would be used. The crux of the issue is in how the work of a computation is divided among the processors: it might seem at first that the best way to divide a computation among several processors is to give each a different part of the program to execute. This works to an extent, but (as the aforementioned journalistic analogy suggests) it suffers from the same drawbacks as assigning a task to a team of people: much of the potential concurrency is lost in the problems associated with coordination. Programming a computer by breaking up the program is like having a large team of people paint a fence by assigning one person to the job of opening the paint, another to preparing the surface, another to applying the paint, and another to cleaning the brushes. This functional breakdown requires a high degree of coordination, and after a certain point adding more people doesn't help speed up the task.

A more efficient way to use a parallel computer is to have each processor perform similar work, but on a different section of the data. This so-called *data parallel decomposition* of the task is like getting a fence painted by assigning each worker a separate section of fence. Not all problems break up as easily as painting a fence, but where large compuations are concerned this method works surprisingly well. For instance, image-processing tasks can be composed in concurrent parts by assigning a little patch of the image to each processor. A search problem, like playing chess, can be decomposed by having each processor simultaneously search different lines of play. In these examples, the degree of speed-up achieved is almost proportional to the number of processors—so the more of them the better. A little additional time must be spent dividing the problem among the processors and gluing the answers together, but if the problem is large the computation can be performed very efficiently, even on a parallel machine with tens of thousands of processors.

The computations just described can fairly obviously be decomposed to run in parallel, but data parallel decomposition also works on more complicated tasks. There are surprisingly few large problems that cannot be handled efficiently by parallel processing. Even problems that most people think are inherently sequential can usually be solved efficiently on a parallel computer. An example is the *chain-following problem*. My children play a game called Treasure Hunt, which is based on the chain-following problem. I give them a piece of paper with a clue about where the next clue is hidden. That clue leads to the next clue, and so on, until they reach the treasure at the end. In the computational version of this game, the program is given, as input, the address of a location in memory containing the address of another location. That location contains the address of still another, and so on. Eventually, the address specifies a memory location containing a special word that indicates it is the end of the chain. The problem is to find the last location from the first.

Initially, the chain-following problem looks like the epitome of an inherently sequential computation, because there seems to be no way for the computer to find the last location in the chain without following the linked addresses through the entire chain. The computer has to look at the first location to find the address of the second, then look at the second to find the address of the third, and so on. It turns out, however, that the problem can be solved in parallel. A parallel computer with a million processors can find the last element in a chain of a million addresses in twenty steps.

The trick is to divide the problem in half at every step, a bit like the approach used in the sorting algorithms in chapter 5. Assume that each of the million memory locations has its own processor, which can send a message to any other processor. To find the end of the chain, every processor begins by sending its own address to the processor that follows it in the chain—that is, the processor whose address is stored in its memory location. Each processor then knows the address not only of the processor that comes after it but also of the one that precedes it. The processor uses this information to send the address of its successor to its predecessor. Now each processor knows the address of the processor that lies two ahead of it in the chain, so now the chain connecting the first processor to the last processor is half the length it was originally. This reduction step is then repeated, and each time it is repeated, the length of the chain is halved. After twenty iterations of the reduction step, the first processor in a million-memory chain knows the address of the last. Similar methods can be applied to many other tasks that seem inherently sequential.

As of this writing, parallel computers are still relatively new, and it is not yet understood what other types of problems can be decomposed to efficiently take advantage of many processors. A rule of thumb seems to be that parallelism works best on problems with large amounts of data, because if there are lots of data elements there is plenty of similar work to divide among processors.

One reason that most computations can be decomposed into concurrent subproblems is that most computations are models of the physical world. Computations based on physical models can operate in parallel because the physical world operates in parallel. Computer graphics images, for example, are often synthesized by an algorithm that simulates the physical process of light reflecting off the surfaces of objects. The picture is drawn from a mathematical description of the shapes of the objects by calculating how each ray of light would bounce from surface to surface while traveling from the source to the eye. The calculation of all the light rays can proceed concurrently, because the bouncing of light proceeds concurrently in the physical world.

A typical example of the kind of computation well suited to a parallel computer is a simulation of the atmosphere, used in predicting the weather. The three-dimensional array of numbers that represents the atmosphere is analogous to three-dimensional physical space. Each number specifies the physical parameter of a certain volume of air—for example, the pressure in a cube of atmosphere 1 kilometer on a side. Each of these cubes will be represented by a few numbers specifying average temperature, pressure, wind velocity, and humidity. To predict how the volume of air in one such cube will evolve, the computer calculates how air flows between neighboring volumes: for instance, if more air flows into a volume than out of it, then the pressure in that volume will increase. The computer also calculates the changes attributable to factors such as sunlight and evaporation. The atmospheric simulation is calculated in a series of steps, each of which corresponds to a small increment of time—say, half an hour—so that the flow of simulated air and water among the cells in the array is analogous to the flow of real air and water in the pattern of weather. The result is a kind of three-dimensional moving picture inside the computer—a picture that behaves according to physical laws.

Of course, the accuracy of this simulation will depend both on the resolution and the accuracy of the three-dimensional image, which accounts for the notorious inaccuracy of weather predictions over time. If the resolution of the model is increased and the initial conditions are measured more accurately, then the prediction will be better—although even a very high resolution will never be perfect over a long period of time, because the initial state of the atmosphere cannot be measured with exact precision. Like the game of roulette, weather systems are chaotic, so a small change in initial conditions will produce a significant change in the outcome. On a parallel computer, each processor can be assigned the responsibility for predicting the weather in a tiny area. When wind blows from one area to the next, then the processors modeling those areas must communicate. Processors modeling geographically separated areas can proceed almost independently, in parallel, because the weather in these areas is almost independent. The computation is local and concurrent, because the physics that governs the weather is also local and concurrent.

While the weather simulation is linked to physical law in an obvious manner, many other computations are linked more subtly to the physical world. For instance, calculating telephone bills is concurrent, because telephones (and telephone customers) operate independently in the physical world. The only problems we don't know how to solve efficiently on a parallel computer are those for which the growing dimension of the problem is analogous to the passage of time. An example is the problem of predicting the future positions of the planets. (Ironically, this is the very problem for which many of our mathematical tools of computation were originally invented.)

The paths of the planets are the consequence of well-defined rules of momentum and gravitational interactions between the nine planets and the Sun. (For simplicity's sake, we will ignore the effects of small bodies, such as moons and asteroids.) All the information necessary to solve the problem can be represented by nine coordinates, so there isn't much data. The computational difficulty of the problem comes from the fact that the calculation must be made, as far as we know, by calculating the successive positions of the planets at each of billions of tiny steps, each representing a short period of time. The only way we know how to calculate the positions of the planets a million years in the future is to calculate their position at each intermediate time between now and then. If there is a trick to solving this problem concurrently, such as the one used in the chain-following problem, I am not aware of it. On the other hand, as far as I know, no one has proved that this orbit problem is inherently sequential. It remains an open question.

Highly parallel computers are now fairly common. They are used mostly in very large numerical calculations (like the weather simulation) or in large database calculations, such as extracting marketing data from credit card transactions. Since parallel computers are built of the same parts as personal computers, they are likely to become less expensive and more common with time. One of the most interesting parallel computers today is the one that is emerging almost by accident from the networking of sequential machines. The worldwide network of computers called the Internet is still used primarily as a communications system for people. The computers act mostly as a medium—storing and delivering information (like electronic mail) that is meaningful only to humans. I am convinced that this will change. Already standards are beginning to emerge that allow these computers to exchange programs as well as data. The computers on the Internet, working together, have a potential computational capability that far surpasses any individual computer that has ever been constructed.

I believe that eventually the Internet will grow to include the computers embedded in telephone systems, automobiles, and simple home appliances. Such machines will read their inputs directly from the physical world rather than relying on humans as intermediaries. As the information available on the Internet becomes richer, and the types of interaction among the connected computers become more complex, I expect that the Internet will begin to exhibit *emergent behavior* going beyond any that has been explicitly programmed into the system. In fact the Internet is already beginning to show signs of emergent behavior, but so far most of it is pretty simple: plagues of computer viruses and unpredicted patterns of message routing. As computers on the network begin to exchange interacting programs instead of just electronic mail, I suspect that the Internet will start to behave less like a network and more like a parallel computer. I suspect that the emergent behavior of the Internet will get a good deal more interesting.

CHAPTER 8

....

COMPUTERS THAT LEARN AND ADAPT

The computer programs I have so far described operate according to fixed rules supplied by the programmer. They have no way of inventing new rules themselves, or of improving the ones they are given. The chess-playing programs, if their programmers do not tinker with them, will keep making the same mistakes over and over, no matter how many games of chess they play. In this sense, computers are completely predictable; it is in this sense that computers can "do only what they are programmed to do"—a point often raised by the defenders of humankind in the "man vs. machine" debate.

But not all software is this inflexible. It is possible to write programs that improve with experience. When they operate with such programs, computers can learn from their mistakes and correct their own errors. They accomplish this by making use of *feedback*. Any system based on feedback needs three kinds of information:

1. What is the desired state (the *goal*)?
2. What is the difference between the current state and the desired state (the *error*)?
3. What actions will reduce the difference between the current state and the goal state (the *response*).

The feedback system adjusts the response, according to the error, to achieve the goal. The simplest and most familiar examples of feedback systems are not learning systems but control systems; the household thermostat is a good example. This feedback system recognizes only two possible errors and produces only two possible responses. The goal is to maintain a particular temperature, and the possible errors are that the temperature is either too hot or too cold. The responses are predetermined: if the temperature is too cold, the response is to turn on the furnace; if the temperature is too hot, the response is to turn the furnace off. Since the thermostat can only turn the furnace on or off, the response has nothing to do with the magnitude of the error. (This is a fact that I have repeatedly tried to explain to various members of my household, who insist on turning the thermostat all the way up to 90 degrees whenever the house is too cold, in hopes that somehow things will warm up faster; this tactic does not help. The thermostat can only turn the furnace on; it cannot turn it up.)

In principle, however, there is no reason why a home heating thermostat could not respond in proportion to the error. Such a system would require a method of adjusting the output of the furnace, rather than just turning it on or off. The apparatus would doubtless be more complicated and expensive, but it would ensure that the temperature was more precisely maintained. Such *proportional control* thermostats are used today in systems that control delicate industrial processes. Some household appliances—certain models of Japanese washing machines, for example—also use proportional control (or an approximation of it), a feature often advertised as *fuzzy logic*.

Another example of a system that uses proportional control is the automatic-pilot system that guides an airplane. The goal in this case is to keep the plane pointed in a given direction. A direction finder, such as a compass, measures the error in the direction the plane is traveling. The autopilot responds by making an adjustment in the position of the rudder of the plane in proportion to the size and direction of the error. Thus a slight deviation in course will result in only a slight change in the rudder position, but a major deviation, such as the swerving produced by a shift in the wind, will result in a large change in the setting of the rudder. If the automatic pilot system did not use proportional control but instead pushed the rudder all the way to the left or all the way to the right in the fashion of the home thermostat, the airplane would oscillate back and forth in an uncomfortable and probably dangerous manner.

In all of these feedback systems, the connection between error and response is fixed. The sensitivity of the response is predetermined by the design of the control system. But it is also possible to design an even more flexible feedback system, in which the response of the system *adapts* with time. In this case, the parameters of an initial feedback system are adjusted by a second feedback system. If the second feedback system adapts and improves over time, the system can be said to have "learned" the parameters of control.

Consider, for example, how a human pilot learns to fly an airplane. Typically, the student pilot oversteers at first—that is, overcorrects for every error. The pilot is using something like the system the thermostat uses to control the heat: if the plane is too far to the left, turn right; if it is too far to the right, turn left. Since there is a delay between turning the control rudder and the response of the plane, the system begins to oscillate. The pilot needs to learn how to move the rudder in proportion to the error, and this requires gauging the sensitivity of the response. The pilot learns this parameter through another feedback system; in this case, the *goal* of the feedback is to keep the plane on the correct course without oscillations, and the *error* is the degree of oscillation. The *response* is to adjust the response of the primary feedback system—that is, to adjust the amount that the control rudder is moved in order to correct a given erroneous angle in the plane's heading. Whenever the pilot's first feedback system is oscillating, he reduces its responsiveness. He increases its responsiveness if the plane begins to drift off course. Once the pilot learns the correct sensitivity, he can keep the plane on course without any oscillation.

It would be possible to build an automatic pilot that uses a second feedback system to adjust its own parameters, as described here. In this case, the autopilot could be said to have "learned" to fly the plane, in the same way that the human pilot learns. As far as I know, such adaptive autopilot systems are not used in real airplanes, but if they were they would have certain advantages. If the airplane sustained damage that caused a change in the responsiveness of the plane, such as a partly broken rudder, the autopilot would be able to adapt to this new situation. It might even be able to adapt if the connections to the rudder's control motor were accidentally reversed, so that the signal that normally turns the airplane right instead turned it left. Like a human pilot, the autopilot would require a fair amount of time to adjust to such a radical change in circumstances.

TRAINING THE COMPUTER

This basic notion of feedback is central to all learning systems, although it often takes a more complicated form than the self-adjusting automatic pilot. Often, feedback in computer programs is provided by *training* with the help of examples. The trainer (usually a human being) plays the role of a teacher, and the program becomes a student. A classical example of a trained learning system is a program written by the AI pioneer Patrick Winston, which learns the definition of concepts like "arch" from a series of positive and negative examples provided by an instructor. Winston's program learns new concepts by looking at simple line drawings of piles of blocks. The program is able to analyze such drawings and generate symbolic descriptions of the piles of blocks: for example, "Two touching cubes, supporting a wedge." The trainer shows the program some examples of block configurations that form arches and another set of examples that do not, telling the program which are examples of "arch" and which are not. Initially, the program has no definition for the concept of "arch," but as it is shown these positive and negative examples, it begins to formulate a working definition. Each time the program is shown a new example, it tests its working definition against the new example. If the definition sufficiently describes a positive example, or rules out a negative example, the program does not modify the definition. If the definition is in error, it is modified to fit the example.

Here is a scenario of how the program learns the definition of "arch" from a few examples. Assume that the first example the program is shown is a positive example: example A in Figure 25, two upright rectangular blocks supporting a triangle. To start, the program will have to make an initial guess at formulating the definition of an arch. This initial guess does not need to be accurate, because it will be refined by future examples. Let's assume that the program uses the shapes of the blocks as its initial guess at a definition: "An arch is two rectangular blocks and a triangular block." The second example the program is shown might be the same blocks, all lying down (example C in Figure 25). This is a negative example—that is, an example of something that is *not* an arch. Since the program's initial working definition mistakenly identifies this negative example as an arch, it will modify its definition to exclude the example. The program does this by identifying differences between the defini-

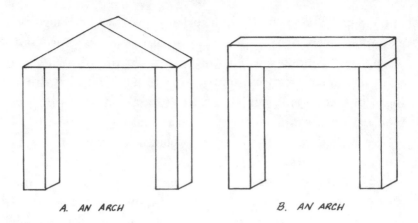

A. AN ARCH   B. AN ARCH

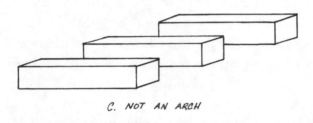

C. NOT AN ARCH

FIGURE 25

Positive and negative examples of arches

tion and the example and using them to add restrictions to the definition. In this case, the difference is in the relationships of the blocks, so an improved definition will include these relationships: "An arch is two *upright* rectangular blocks *supporting* a triangular block." Now let's say the trainer supplies another positive example (B in Figure 25). This example uses a rectangular block at the top, instead of a triangular one. Since the program's working definition is not broad enough to include this positive example, the program will generalize its definition of "arch" to allow other shapes.

After being shown these examples and a few others, the program will converge on the following definition of an arch: "A prismatic body supported by two upright blocks that do not touch one another." Each element of the definition has been learned by making some kind of mistake, and the definition has been adjusted accordingly. Once the program converges on the right definition, it stops making mistakes and leaves its definition unchanged. It can then correctly identify as an arch any arch it is shown, even if it has never seen that particular set of blocks before. It has learned the concept of "arch."

NEURAL NETWORKS

Winston's program learns the concept of "arch," but concepts like "touching," "triangular block," and "support" have been built into it from the beginning. Its representation of the world is specifically designed for piles of blocks. The search for a more general, universal representation scheme has led many researchers to computing systems with structures analogous to connected nets of biological neurons, such as occur in the brain. Such a system is called an artificial *neural network*.

A neural network is a simulated network of artificial neurons. This simulation may be performed on any kind of computer, but because the artificial neurons can operate concurrently, a parallel computer is the most natural place to execute it. Each artificial neuron has one output and a large number of inputs, perhaps hundreds or thousands. In the most common type of neural network, the signals between the neurons are binary—that is, either 1 or 0. The output of one neuron can be connected to the inputs of many others. Each input has a number associated with it, called its *weight*, which determines how much of an effect the input has upon the neuron's single output. This weight can be any number, positive or negative. The neuron's output is thus determined by a vote of the signals coming into its inputs, adjusted by the weights of the inputs. The neuron computes its output by multiplying each input signal by the input weight and summing the results; in other words, it adds up the weight of all the inputs that receive a 1 signal. If the weighted sum reaches a specific threshold, the output is 1; otherwise, the output is 0.

The function of an artificial neuron corresponds, very roughly, to the function of some types of real neurons in the brain. Real neurons also have one output and many inputs, and the input connections, called *synapses,* have different strengths (corresponding to the different input weights). A signal can either enhance or inhibit the firing of the neuron (corresponding to positive and negative weights), and the neuron will fire when the combined stimulation of the inputs is equal to or above some threshold. These are the senses in which an artificial neuron is analogous to a real one. There are also many ways in which a real neuron is much more complicated than an artificial one, but this simple artificial neuron is sufficient for building a system capable of learning.

The first thing to notice about artificial neurons is that they can be used to carry out the **And**, **Or**, and **Invert** operations. A neuron implements the **Or** function if the threshold is 1 and each of the input weights is equal to or greater than 1. A neuron with a threshold equal to the sum of the weights will implement the **And** function. Neurons with a single, negatively weighted input and a threshold of 0 will implement the **Invert** function. Since any logical function can be constructed by combining the **And**, **Or,** and **Invert** functions, a network of neurons can implement any Boolean function. Artificial neurons are universal building blocks.

We don't know very much about how the human brain works, but in some parts of the brain it seems that new information is learned by modifying the strength of the synapses that connect the neurons. This is certainly the case in the lower organisms on which we perform experiments—for example, sea snails. Sea snails can be taught certain conditioned responses, and it can be shown that they learn the response by changing the strength of the synaptic connections between neurons. Assuming that human learning works the same way, you are (I hope) adjusting the connections in your brain as you read this book.

A network of artificial neurons can "learn" by changing the weights of its connections. A good example is a very simple type of neural network called a *perceptron*, which can learn to recognize patterns. The way perceptrons learn is indicative of how most neural networks operate. A perceptron is a network with two layers of neurons and a single output. Each input in the first layer is connected to a sensing device like a light detector, which measures the brightness of one spot on an image. Each input of the second layer is connected to an output from the first layer, as shown in Figure 26.

Imagine that we are trying to teach the perceptron to recognize the letter A, which we will accomplish by showing it a large number of positive and negative examples of an A. The goal is for the perceptron to adjust the weights of the second layer so that its output will be 1 if, and only if, it is shown the image of an A. It accomplishes this by adjusting those weights whenever it makes an error. Each neuron in the first layer of the perceptron looks at a small patch of whatever example is being presented. Each of these first-layer neurons is programmed to recognize a specific local feature, such as a particular corner or a line at a particular orientation; it does so by means of the fixed weights of its own inputs. For example, here is a pattern of negative and positive input weights for the receptive field of a first-layer neuron programmed to recognize a corner, such as the point at the top of a capital A:

```
  -         -         -         -         -         -
  -         -         +         -         -         -
  -         +         +         +         -         -
  -         +         +         +         +         -
  +         +         +         +         +         +
  +         +         +         +         +         +
```

The first layer of the perceptron contains thousands of such feature-detecting neurons, each one programmed to recognize a particular kind of local feature in a particular part of the receptive field. This first layer of neurons detects features in the image which are useful for distinguishing between any letters; serifs are easy to detect, so they make

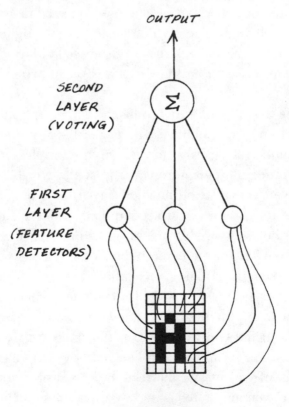

FIGURE 26

Perceptron

letters more recognizable to the perceptron, just as they make a particular letter easier for the human eye to identify.

The local-feature detectors in the first layer provide the evidence, and the weights of the second layer determine how to weigh this evidence. For example, a corner pointing upward in the upper part of the image is evidence in favor of an A, while a corner pointing downward in the middle is evidence against. The perceptron learns by adjusting the weights on the inputs to the second layer. The learning algorithm is very simple: whenever the trainer indicates that the perceptron has made a mistake, the perceptron will adjust all of the weights of all the inputs that voted in favor of the mistake in such a way as to make future mistakes less likely. For instance, if the perceptron incorrectly identifies an image as an A, the weights of all the inputs that voted in favor of the false conclusion will be decreased. If the perceptron fails to identify a real A, then the inputs that voted in favor of the A will be increased. If the perceptron has enough feature detectors of the right type, this training method will eventually cause the perceptron to learn to recognize A's.

The learning procedure of the perceptron is another example of feedback. The *goal* is to set the weights correctly, the *errors* are misidentifications of the training examples, and the *response* is to adjust the weights. Notice that perceptrons, like Winston's arch program, learn only by making mistakes. This is a characteristic of all feedback-based learning systems. Given enough training, this particular procedure will always converge upon a correct choice of weights, assuming that there is a set of weights that does the job. This makes the perceptron seem like the perfect pattern-recognition machine, but the catch is the assumption that there exists some correct pattern of weights that will accomplish the task. To recognize the letter A in various sizes, fonts, and positions, the perceptron needs a very rich set of feature detectors in the first layer.

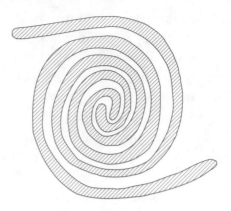

FIGURE 27

Perceptron spiral

Perceptrons can learn to recognize any letter if they are given enough features to work with, but there are some types of patterns, more complex than letters, that cannot be recognized by summing together local features in any way. For example, simply by summing up the evidence of local patches, a perceptron cannot tell whether or not all the dark spots in an image are connected, because connectedness is a global property; no local feature, by itself, can serve as evidence for or against connectedness. Figure 27, adapted from Marvin Minsky and Seymour Papert's book *Perceptrons*, demonstrates that connectedness cannot always be assessed just by looking at local features.

For these and other reasons, two-layer perceptrons are not the most practical neural networks for recognizing most types of patterns. More general neural networks, with more layers, are able to recognize more complicated patterns. Such networks use similar procedures for learning. Trained neural networks of this type are often used for tasks like image recognition and speech recognition—tasks that are difficult to specify by a fixed set of rules. For instance, the simple word-recognition systems that are built into many children's toys today are based on neural networks.

SELF-ORGANIZING SYSTEMS
....

The disadvantage of a learning system based on positive and negative examples is that it requires a trainer to classify the examples. There is another type of neural network, which does not require a trainer—or, to put it another way, there are networks in which the training signals are generated by the network itself. Such a self-training network is a *self-organizing* system. Self-organizing systems have been studied for years (Alan Turing published important work in this area), but there has been a recent renewal of research activity in such systems, and even some new progress, partly because of the availability of faster computers. Like trained neural networks, self-organizing systems are a natural fit for parallel computers.

As an example of a self-organizing system that works, consider the problem of transmitting an image from the eye to the brain (see Figure 28). The retina, on which the image is projected, is a two-dimensional sheet of light-sensitive neurons. The image on the retina is transformed into a similar projected image in the brain by a bundle of neurons that transfers the image. If this bundle is wired imperfectly, then the projected image will be slightly scrambled, with each pixel in slightly the wrong place. I will describe a self-organizing artificial neural network that can learn to unscramble such a picture, restoring each pixel to its proper position. The unscrambler consists of a single layer of neurons arranged in a two-dimensional array. The outputs of these neurons form the corrected image. If the picture is scrambled only slightly, then each pixel in the scrambled image will be in the general neighborhood of its correct position. Each neuron's inputs look at a neighborhood of pixels in the scrambled image, and the neuron learns which of these pixels should be connected to the output in order to produce the unscrambled image. The neuron forms the connection by setting the weight of the correct input to 1 and the weight of its other inputs to 0.

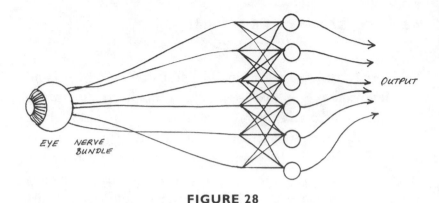

EYE   NERVE
       BUNDLE

OUTPUT

FIGURE 28

Eye, with scrambled nerve bundle and unscrambler

The training algorithm for the unscrambler is based on the fact that images have a nonrandom structure. As discussed earlier, real images are not just random arrays of dots but pictures of the world, so nearby areas of the image tend to look the same. The unscrambler turns this statement around, by assuming that pixels that tend to look the same ought therefore to be near one another. The neurons in the unscrambler work by measuring the correlation of each of their inputs with the outputs of the neighboring neurons during exposure to a series of images. Whenever a neuron makes an "error" by firing differently from its neighbors, the neuron increases the weight of the inputs that match the outputs of its neighbors and decreases the weights of the other inputs. Of course, its neighbors are also learning their connections at the same time, so in the beginning it is a case of (so to speak) the blind leading the blind, but eventually some of the unscrambler neurons begin locking onto their correct inputs and thus become effective trainers for their neighbors. Again, the only neurons that are adjusted are those that have made mistakes. As the neurons train one another, an unscrambled image begins to emerge in the outputs, and eventually the network organizes itself to produce an image of perfect clarity.

The self-adjusting autopilot, Patrick Winston's arch program, the perceptron, and the unscrambler are just a few examples of systems that learn. All these systems are based on either external or internal feedbacks, and all learn by correcting their mistakes. The design of each of these systems was inspired by a biological system of similar function. In harvesting these products of evolution, we are like the fool in Aesop's fable, "The Goose That Laid the Golden Egg," who chooses the eggs instead of the goose. In the next chapter, we shall discuss the goose.

....

CHAPTER 9

BEYOND ENGINEERING

According to legend, the thirteenth-century scientist and monk Roger Bacon was a dabbler in black magic, and once constructed a talking mechanical head. It is said that he wanted to defend England from invaders by building a wall around the kingdom, and he constructed the head in order to ask its advice about how to build the wall. Bacon fashioned the head out of brass, replicating the design of a human head in every detail. He heated it over a fire while uttering magical incantations—a process that went on for days. Eventually, the head awoke and began to talk. Unfortunately, Bacon was by that time so exhausted from casting spells that he had fallen asleep. His young assistant was unwilling to awaken the master for the mere ramblings of a brass head, and the head exploded over the fire before Bacon could ask it any questions.

The Bacon legend has elements in common with stories about other conjurers who constructed an artificial intelligence: Dedalus, Pygmalion, Albertus Magnus, the Rabbi of Prague. A theme common to many of these stories is that some form of cooking or ripening is necessary to make something start to think. In the days before computing machines, few imagined that a process as complex as thinking could ever be broken down into operations that could be implemented by mechanisms. Instead, the assumption was that if an intelligence were ever to be created, it would be by an emergent process—that is, by a process in which the complex behavior emerges as a global consequence of billions of tiny local interactions. It was assumed that what the conjurer needed was not the correct wiring diagram but the correct recipe, according to which the ingredients would organize *themselves* into an intelligence. Such a process would allow an intelligence to be created without the conjurer's understanding exactly how the process—or the intelligence itself—worked.

Oddly enough, I am in basic agreement with this prescientific notion: I believe that we may be able create an artificial intelligence long before we understand natural intelligence, and I suspect that the creation process will be one in which we arrange for intelligence to emerge from a complex series of interactions that we do not understand in detail—that is, a process less like engineering a machine and more like baking a cake or growing a garden. We will not engineer an artificial intelligence; rather, we will set up the right conditions under which an intelligence can emerge. The greatest achievement of our technology may well be the creation of tools that allow us to go *beyond* engineering— that allow us to create more than we can understand.

Before discussing how this emergent design process might work, let us consider our best example of intelligence: the human brain. Since the brain itself was "designed"—by the emergent process of Darwinian evolution—it may be usefully compared with the engineered designs that we have considered so far.

THE BRAIN

The human brain has about 10^{12} neurons, and each neuron has, on average, 10^5 connections. The brain is to some degree a self-organizing system, but it would be wrong to think of it as a homogeneous mass. It contains hundreds of different types of neurons, many of which occur only in particular regions. Studies of brain tissue show that the patterns of neuronal connection, too, differ in the various regions of the brain: there are some fifty areas in which the pattern is recognizably different, and there are probably many more in which the differences in neural anatomy are too subtle for us to distinguish.

Each area of the brain is apparently specialized for a particular type of function, such as recognizing color in visual images, producing intonation in speech, or keeping track of the names of things. We know this because when specific areas are damaged by an accident or a stroke there is also a corresponding loss of function. For example, damage to areas 44 and 45 on the left side of the frontal lobe—together they are called Broca's area—often robs someone of the ability to produce grammatical speech. People so afflicted may still pronounce words clearly and they may understand the speech of others, but they will be unable to construct grammatical sentences. Damage to an area known as the annular gyrus, located a little farther toward the back of the head, causes difficulties in reading and writing; damage to yet other areas results in an inability to recall the names of familiar objects or recognize familiar faces.

It would be wrong to assume that the various areas of the brain are analogous to the functional building blocks of a computer. For one thing, damage to most areas will not cause a well-defined loss of function: removal of most of the right frontal lobe, for example, sometimes causes indefinable changes in personality and sometimes causes no noticeable change at all. Even in those cases in which the loss of function is well-defined, it is not at all evident that the function was performed entirely by the damaged area; it may be that the area just provided some minor element of support necessary for the function. An automobile with a dead battery will not be able to move, but we don't therefore assume that the battery is responsible for propelling the car.

There are certain areas of the brain—in particular, areas near the back of the head associated with visual processing— where we can actually make some sense of the pattern of connections: for instance, those involved in receiving the inputs from the left and right eye to create the sense of depth in stereo vision. But in most of the brain, the "wiring pattern" remains a mystery. Even the notion that most of the brain is hardwired for specific functions may turn out to be incorrect. Language, for instance, seems to be processed mostly on the left side, whereas spatial recognition, such as the ability to understand a map, seems to be performed primarily on the right. Yet under a microscope the pattern of left-brain and right-brain tissue looks pretty much the same. If there is a systematic difference between the wiring patterns in the brain's two hemispheres, it is too subtle for us to discern.

It may be that brain functions are learned in some sort of self-organizing process that changes the strength of various synaptic connections in order to fit an area for a certain kind of function. This is surely true to a degree. We know, for instance, that a monkey with a missing finger will continue to use the area of its brain that normally processes information from that finger: the idle neurons are recruited to perform processing for the animal's other fingers. Human beings probably rearrange the functions of the brain in a similar manner as they recover from a stroke. A stroke victim may initially have trouble with a specific function, such as recognizing faces, and then relearn the function with time. Since damaged neurons cannot regenerate themselves, the patient presumably relearns the function by recruiting neurons in a different part of the brain.

If functions like recognizing faces and understanding language are learned in different parts of the brain, then there must also be some sense in which these functions are already built in from the beginning. Newborn babies are particularly interested in faces in the first few days of life, and they learn to recognize them long before they learn to distinguish between much simpler shapes, such as letters. Similarly, babies seem predisposed to pay attention to certain kinds of patterns in speech that allow them to learn words and grammar. The functions that process language and recognize faces end up in different parts of the brain because, presumably, those parts of the brain are somehow primed to perform those dissimilar functions.

Even in those portions of the brain where functions seem to be hardwired, the pattern of wiring bears little resemblance to the hierarchical structure of functional blocks within a computer: there is no simple pattern of inputs going to outputs. Instead the connections are often bidirectional, with one set of neurons connecting in one direction and a complementary set connecting in reverse. Figure 29 shows the wiring diagram of the visual cortex of the macaque monkey, as best as can be determined by tracing the connections. Each of the lines in the diagram represents a bundle of many thousands of neurons, along with a complementary bundle in the reverse direction. At first glance, it appears as if everything is connected to everything else—unlike the neat, hierarchical circuit diagram of an engineered computer.

The important point here is that the brain is not only very complicated but also very different in structure from an engineered machine. That does not mean that we cannot ever engineer a machine to perform the functions of the human brain, but it does mean that we cannot expect to understand an intelligence by taking it apart and analyzing it as if it were a hierarchically designed machine.

It is possible that a satisfactory description of what the brain does will be almost as complex as a description of the structure of the brain—in which case, there is no meaningful sense in which we can understand it. In engineering, the way we deal with complexity is to break it into parts. Once we

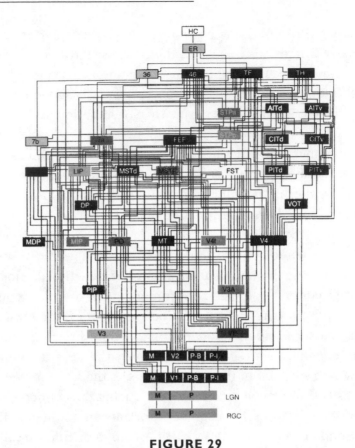

FIGURE 29

Block diagram of the macaque visual cortex

understand each part separately, we can understand the interactions between the parts. The way we understand each of
the parts is to apply the engineering process recursively,
breaking each part into a subpart, and so on. The design of an
electronic computer, along with all its software, is impressive
testimony to how far this process can be pushed. As long as
the function of each part is carefully specified and implemented, and as long as the interactions between the parts are
controlled and predictable, this system of "divide and con-
quer" works very well, but an evolved object like the brain
does not necessarily have this kind of hierarchical structure.

THE PROBLEM WITH MODULARITY
....

The reliance on a strict hierarchical structure is the Achilles heel of the engineering process, since of necessity it creates the kind of adamant inflexibility we associate with machines. As discussed in chapter 6, hierarchical systems are fragile in the sense that they are prone to catastrophic failure. Products of engineering are inherently fragile, because each part of an engineered system must meet the design specifications of how it should interact with other parts. These specifications serve as a kind of contract between components. If one of the components breaks its part of the contract, the design assumptions of the systems are invalid, and the system breaks down in an unpredictable way. The failure of a single low-level component can percolate through the system with catastrophic effects. Of course, complex systems like computers and airplanes are engineered to avoid these so-called single-point failures, through the methods of redundancy described in chapter 6, but such techniques can guard the system only against anticipated failures. All the potential consequences of a particular failure must be predicted and understood—a task that becomes increasingly difficult as the machine becomes more and more complex.

The problem goes beyond the failure of individual components. In a complicated system, even correctly functioning parts can produce unexpected behaviors when they interact. Often when a large software system malfunctions, the programmers responsible for each of the parts can convincingly argue that each of their respective subroutines is doing the right thing. Often they are all correct, in the sense that each subroutine is correctly implementing its own specified function. The flaw lies in the specifications of what the parts are supposed to do and how they are supposed to interact. Such specifications are difficult to write correctly without anticipating all possible interactions. Large complex systems, like computer operating systems or telephone networks, often exhibit puzzling and unanticipated behaviors even when every part is functioning as designed. You may recall that a few years ago the long-distance telephone lines of the eastern United States stopped routing calls for several hours. The system used a sophisticated fault-tolerant design, based on redundancy. All its components were functioning correctly, but an unanticipated interaction between two versions of the software running at different switching stations caused the entire system to fail.

It is amazing to me that the engineering process works as well as it does. Designing something as complicated as a computer or an operating system can require thousands of people. If the system is sufficiently complicated, no one person can have a complete view of the system. This situation generally leads to mistakes stemming from misunderstandings of interfaces and inefficiencies of design. Again, such interface difficulties get worse as the system becomes more complex.

It is important to note that the problems outlined above are not inherent weakness of machines or of software per se. They are weaknesses of the engineering design process. We know that not everything that is complex is fragile. The brain is much more complicated than a computer, yet it is much less prone to catastrophic failure. The contrast in reliability between the brain and the computer illustrates the difference between the products of evolution and those of engineering. A single error in a computer's program can cause it to crash, but the brain is usually able to tolerate bad ideas and incorrect information and even malfunctioning components. Individual neurons in the brain are constantly dying, and are never replaced; unless the damage is severe, the brain manages to adapt and compensate for these failures. (Ironically, as I was writing this chapter, my computer crashed and required rebooting.) Humans rarely crash.

SIMULATED EVOLUTION
••••••••••••••••••••••••••••••

So, in creating an artificial intelligence, what is the alternative to engineering? One approach is to mimic within the computer the process of biological evolution. Simulated evolution gives us a different way to design complicated hardware and software—a way that avoids many of the problems of engineering. To understand how simulated evolution works, let's look at a specific example. Say that we want to design a piece of software that sorts numbers into descending order. The standard engineering approach would be to write such a program using one of the sorting algorithms discussed in chapter 5, but let's consider how we might instead "evolve" the software.

The first step is to generate a "population" of random programs. We can create this population using a pseudorandom number generator to produce random sequences of instructions (see chapter 4). To speed up the process, we can use only those instructions useful for sorting, such as comparison and exchange instructions. Each of these random sequences of instructions is a program: the random population will contain, say, 10,000 such programs, each one a few hundred instructions long.

The next step is to test the population to find which programs are the most successful. This requires us to run each of the programs to see whether or not it can sort a test sequence correctly. Of course, since the programs are random, none are likely to pass the test—but by sheer luck some will come closer to a correct sorting than others. For instance, by chance, a program may move low numbers to the back of the sequence. By testing each program on a few different number sequences, we can assign a fitness score to each program.

The next step is to create new populations descended from the high-scoring programs. To accomplish this, programs with less than average scores are deleted; only the fittest programs survive. The new population is created by making copies of the surviving programs with minor random variations, a process analogous to asexual reproduction with mutation. Alternatively, we can "breed" new programs by pairing survivors in the previous generation—a process analogous to sexual reproduction. We accomplish this by combining instruction sequences from each of the "parent" programs to produce a "child." The parents presumably survived because they contained useful instruction sequences, and there is a good chance that the child will inherit the most useful traits from each of the parents.

When the new generation of programs is produced, it is again subjected to the same testing and selection procedure, so that once again the fittest programs survive and reproduce. A parallel computer will produce a new generation every few seconds, so the selection and variation processes can feasibly be repeated many thousands of times. With each generation, the average fitness of the population tends to increase—that is, the programs get better and better at sorting. After a few thousand generations, the programs will sort perfectly.

I have used simulated evolution to evolve a program to solve specific sorting problems, so I know that the process works as described. In my experiments, I also favored the programs that sorted the test sequences quickly, so that faster programs were more likely to survive. This evolutionary process created very fast sorting programs. For the problems I was interested in, the programs that evolved were actually slightly faster than any of the algorithms described in chapter 5—and, in fact, they were faster at sorting numbers than any program I could have written myself.

One of the interesting things about the sorting programs that evolved in my experiment is that I do not understand how they work. I have carefully examined their instruction sequences, but I do not understand them: I have no simpler explanation of how the programs work than the instruction sequences themselves. It may be that the programs are not understandable—that there is no way to break the operation of the program into a hierarchy of understandable parts. If this is true—if evolution can produce something as simple as a sorting program which is fundamentally incomprehensible—it does not bode well for our prospects of ever understanding the human brain.

I have used mathematical tests to prove that the evolved sorting programs are flawless sorters, but I have even more faith in the process that produced them than in the mathematical tests. This is because I know that each of the evolved sorting programs descends from a long line of programs whose survival depended on being able to sort.

The fact that evolved software cannot always be understood makes some people nervous about using it in real applications, but I think this nervousness is founded on false assumptions. One of the assumptions is that engineered systems are always well understood, but this is true only of relatively simple systems. As noted, no single person completely understands a complex operating system. The second false assumption is that systems are less trustworthy if they cannot be explained. Given the choice of flying in an airplane operated by an engineered computer program or one flown by a human pilot, I would pick the human pilot. And I would do so even though I don't understand how the human pilot works. I prefer to put my faith in the process that produced the pilot. As with the sorting programs, I know that a pilot is descended from a long line of survivors. If the safety of the airplane depended on sorting numbers correctly, I would rather depend on an evolved sorting program than on one written by a team of programmers.

EVOLVING A THINKING MACHINE

Simulated evolution is not in itself a solution to the problem of making a thinking machine, but it points us in the right direction. The key idea is to shift the burden of complexity away from the hierarchy of design and onto the combinatorial power of the computer. Essentially, simulated evolution is a kind of heuristic search technique that searches the space of possible designs. The heuristics it uses to search the space are *Try a design similar to the best designs you have found so far* and *Combine elements of two successful designs.* Both heuristics work well.

Simulated evolution is a good way to create novel structures, but it is an inefficient way to tune an existing design. Its weaknesses as well as its strengths stem from evolution's inherent blindness to the "Why" of a design. Unlike the feedback systems described in the last chapter, where specific changes were made to correct specific failures, evolution chooses variations blindly, without taking into account how the changes will affect the outcome.

The human brain takes advantage of both mechanisms: it is as much a product of learning as it is of evolution. Evolution paints the broad strokes, and the development of the individual in interaction with its environment completes the picture. In fact, the product of evolution is not so much a design for a brain as the design for a process that generates a brain—not so much a blueprint as a recipe. Thus, there are multiple levels of emergent processes operating at once. An evolutionary process creates a recipe for growing a brain, and the developmental process interacts with the environment to wire the brain. The developmental process includes both the internally driven processes of morphogenesis and the externally driven processes of learning. The maturational forces of morphogenesis cause nerve cells to grow in the right patterns, and the process of learning fine-tunes the connections. The ultimate stage in the brain's learning is a cultural process, in which knowledge acquired by other individuals over many generations is transferred into it.

I have described each of these emergent mechanisms (evolution, morphogenesis, learning) as if they were discrete processes, but in reality they are synergistically intertwined. There is no hard line between the maturational forces of morphogenesis and the instructional processes of culture. When a mother coos baby talk to her newborn child, this is both an instruction process and an aid in the maturation of the infant brain. The process of morphogenesis is itself an adaptive process, in which each cell develops in constant interaction with the rest of the cells in the organism, in a complex feedback process that tends to correct errors and keep the development of the organism on track.

There are also synergistic interactions between the evolutionary processes that create the species and the developmental processes that create the individual. The clearest example of the interaction between development and evolution is known as the Baldwin effect, first described by the evolutionary biologist James Baldwin in 1896 and rediscovered by the computer scientist Geoffrey Hinton almost a century later. The basic idea of the Baldwin effect is that when you combine evolution with development, evolution can happen faster; the adaptive processes of development can fix the flaws in an imperfect evolutionary design.

To understand the Baldwin effect, one must first appreciate the difficulty of evolving traits that require multiple mutations to occur together. Consider the evolution of the instinct for nest-building behavior in a bird. It is reasonable to assume that building a nest requires a few dozen individual steps, such as locating a twig, picking it up with the beak, carrying the twig back to the nest, and so on. Let's also assume, for the sake of the example, that each of these steps requires a different mutation and that the benefit to the bird (in the form of a completed nest) requires the complete set of mutations. In other words, if even a single step is missing, the nest will not get built at all, and therefore the bird will be no more fit than its peers and will derive no evolutionary advantage. Obviously, the problem with evolving such a trait is that evolution will select for one of its component mutations only if all the others are present: the simultaneous occurrence of all these mutations within a single individual is a highly improbable event. Since no single step is beneficial by itself, it is difficult to imagine how a behavior such as nest building could possibly evolve.

The Baldwin effect is synergistic interaction between evolution and learning. This interaction helps to solve this problem by offering the bird partial credit for a mutation that produces a single step of the task. A bird that is born knowing how to do some of the steps will have an advantage over a bird that does not, since it will have fewer steps to learn, so it's more likely to arrive at succesful nest-building behavior. Each single step that the bird is born with contributes to the possibility of learning, and therefore is valuable in itself. Viewed this way, each individual mutation will be favored independently, so that nest-building behavior will result from steps that are added to the bird's instinctual repertoire gradually, and in less time than it would take for a probabilistic fluke that produces the mutations all at once in a single individual In effect, the fact that the bird can learn makes the evolution happen faster. The Baldwin effect applies not just to learning but to any adaptive mechanism in the development of the individual.

Part of the reason that I'm optimistic about the prospects of evolving a thinking machine is that we do not have to start from scratch. We can "prime" the initial population of machines with patterns of structure that we observe in the brain. We can also start with whatever patterns of development and learning we observe in natural systems, even if we do not have a complete understanding of them. This should help even if our guesses are not quite right, since starting our search somewhere near a solution is probably much better than starting at random. By including *some* model of development in this process, the evolution of a thinking machine could take advantage of the Baldwin effect.

Another effect that radically reduces the time required to develop a complex behavior is instruction. A human baby develops intelligence at least in part because it has other humans to learn from. Part of this learning is acquired by sheer imitation, and part through explicit instruction. Human language is a spectacular mechanism for transferring ideas from one mind to another, allowing us to accumulate useful knowledge and behavior over many generations at a rate that far outpaces biological evolution. The "recipe" for human intelligence lies as much in human culture as it does in the human genome.

However, even starting with everything we know, I would not expect us to be able to evolve high-level artificial intelligence in a single step. Here is a rough outline of how the sequence of stages might progress. We would begin by evolving a design of a machine with the intelligence of, say, an insect by creating a simple environment in which insectlike intelligence would be favored, and by starting with an initial population predisposed through its developmental mechanisms to develop the kinds of neural structures we see in insects. Through a sequence of successively richer simulated environments, we might eventually evolve our insect intelligence into the intelligence of a frog, a mouse, and so on. Even going this far would doubtless take decades of work and involve many dead ends and false starts, but eventually this course of research could lead to the evolution of an artificial intelligence with the complexity and flexibility of the primate brain.

Should we ever manage to evolve a machine that can understand language, we would be able to skip ahead rapidly, by taking advantage of human culture. I imagine that we would need to teach an intelligent machine by much the same process that we would teach a human child, with the same mixture of skills, facts, morals, and stories. Since we would be incorporating human culture into the machine's recipe for intelligence, the resulting machine would not be an entirely artificial intelligence but rather a human intelligence supported by an artificial mind. For this reason, I expect that we would get along with it just fine.

I am aware, of course, that building such a machine will create a tangle of moral issues. For instance, once such a machine has been created, will it be immoral to turn it off? I would guess that turning it off would be wrong, but I do not pretend to be certain of the moral status of an intelligent artifact. Fortunately, we will have many years to work such questions out.

Most people are interested in not so much the practical moral questions of a hypothetical future as the philosophical issues that the mere possibility an artificial intelligence raises about ourselves. Most of us do not appreciate being likened to machines. This is understandable: we ought to be insulted to be likened to stupid machines, such as toasters and automobiles, or even to today's computers. Saying that the mind is a relative of a current-generation computer is as demeaning as saying that a human being is related to a snail. Yet both statements are true, and both can be helpful. Just as we can learn something about ourselves by studying the neural structure of the snail, we can learn something about ourselves by studying the simple caricature of thought within today's computers. We may be animals, but in a sense our brain is a kind of machine.

Many of my religious friends are shocked that I see the human brain as a machine and the mind as computation. On the other hand, my scientific friends accuse me of being a mystic because I believe that we may never achieve a complete understanding of the phenomenon of thought. Yet I remain convinced that neither religion nor science has everything figured out. I suspect that consciousness is a consequence of the action of normal physical laws, and a manifestation of a complex computation, but to me this makes consciousness no less mysterious and wonderful—if any-

thing, it makes it more so. Between the signals of our neurons and the sensations of our thoughts lies a gap so great
that it may never be bridged by human understanding. So
when I say that the brain is a machine, it is meant not as an
insult to the mind but as an acknowledgment of the potential
of a machine. I do not believe that a human mind is less than
what we imagine it to be, but rather that a machine can be
much, much more.

FURTHER READING

Hillis, W. Daniel. *The Connection Machine.* The MIT Press Series in Artificial Intelligence (MIT, 1989).

Knuth, Donald Ervin. *The Art of Computer Programming,* 4 vols. (Addison-Wesley, 1997).

Minsky, Marvin Lee. *Computation: Finite and Infinite Machines* (Prentice Hall, 1967).

Patterson, David A., and John L. Hennessy. *Computer Architecture: A Quantitative Approach.* 2d ed. (Morgan Kaufman Publishers, 1996).

Weiner, Norbert. *Human Use of Human Beings: Cybernetics and Society* (Avon, 1986).

Winston, Patrick Henry. *Artificial Intelligence.* 3d ed. (Addison-Wesley, 1998).

ACKNOWLEDGMENTS

This project was originally suggested to me by John Brockman, who felt that there was a need for a short book summarizing the ideas behind computers. It seemed to me like a straightforward project at the time, but I soon discovered that writing a short book on such a broad subject can be more difficult than writing a long one.

I thank John and also William Frucht of Basic Books for their help in this endeavor. I wrote the book while visiting at the MIT Media Laboratory, and I would like to thank all of the faculty and students who helped and supported me there, especially the laboratory's founder and director, Nicholas Negroponte. I received significant help in the preparation of the initial manuscript from Debbie Widener, and also from Bettylou McClanahan and Peggi Oakley. Most of what is in the book was taught to me by my friend and mentor Marvin Minsky and other inspiring teachers at MIT, including Gerald Sussman, Claude Shannon, Seymour Papert, Tomaso Poggio, Patrick Winston, and Tom Knight.

I would also like to thank those who read through early drafts and gave me helpful comments: Jerry Lyons, Seymour Papert, George Dyson, Chris Sykes, Brian Eno, Po Bronson, Argye Hillis, and Pati Hillis. I received helpful advice on specific chapters from Tommy Poggio, Neil Gershenfeld, Simon Garfinkel, Mitchell Resnick, and Marvin Minsky. I was very fortunate to have the help of Sara Lippincott, who edited the text, much improving it in the process. Finally, I would like to thank my family: my parents Argye and Bill, who encouraged my interests in designing complex machines; my children, Noah, Asa, and India; and especially my wife, Pati, who patiently encouraged and supported me throughout the project.